Precision Machining Technology Workbook and Projects Manual

Fourth Edition

Peter J. Hoffman
Berks Career & Technology Center, West Campus, Leesport, PA

Eric S. Hopewell
Berks Career & Technology Center, West Campus, Leesport, PA

James Hillwig
Crawford County Career Technical Center, Meadville, PA

Samuel C. Lieber, Ph.D., P.E.
New Jersey Institute of Technology, Newark, NJ

Cengage

Australia • Brazil • Canada • Mexico • Singapore • United Kingdom • United States

Cengage

***Precision Machining Technology Workbook and Projects Manual*, Fourth Edition**

**Peter J. Hoffman/Eric S. Hopewell/
James Hillwig/Samuel C. Lieber**

EVP, General Manager HED: Nhaim Khoury

VP, Portfolio Management: Heather Bradley Cole

Portfolio Director: Laura Stewart

Portfolio Product Manager: Alice Willenbrink

Senior Product Assistant: Jannell Whitted

Learning Operations Manager: Natasha Allen

Senior Learning Designer: Mary Clyne

Content Project Manager:
Yogalakshmi Mohanakrishnan

Senior Manager Product Marketing: Danae April

Product Marketing Manager: Morgan Perin

Production Service: MPS Limited

Senior Designer: Angela Sheehan

Cover image(s): Peter Hoffman/Eric Hopewell

WCN: 01-100-393

Library of Congress Control Number: 2026902337

ISBN: 979-8-214-11417-0

Cengage
10650 Toebben Drive, Suite A
Independence, KY 41051,
USA

Printed at CLDPC, USA, 8-26

Table of Contents

Workbook

Projects

Preface

The *Precision Machining Technology Workbook and Projects Manual* is intended to assist the student in the understanding and development of the fundamental and intermediate machining skills needed for career success in a rapidly changing manufacturing environment.

The Workbook section of the Manual is designed to reinforce the student's understanding and comprehension of the material presented in each unit of the text. After reading and completing each unit of the text, the student should be prepared to successfully complete each corresponding Workbook unit.

The Projects section of the Manual used in conjunction with the text offers a wide range of material to aid in hands-on performance as it relates to NIMS credentialing. At the beginning of each project, there is a list of the various NIMS skills practiced, bringing the student a deeper understanding of that skill as each project is completed. The projects are grouped together so that each NIMS skill can be mastered. Upon successful completion of the projects, the student will be sufficiently prepared to take the corresponding NIMS performance test.

Unit 1 Introduction to Machining

Name: ______________________________ Date: ______________

Score: ______________ Text pages 1–15

Learning Objectives

After completing this unit, you should be able to:

- Define the term *machining*
- Define a machine tool
- Define manufacturing
- List industries that rely on machining
- List manufactured products that rely directly and indirectly on machining
- Compare/contrast manual and CNC machine tools
- Identify the operations that the basic types of machine tools perform

Carefully read the unit, and then answer the following questions.

Multiple Choice

1. The cord of a bow string drill was used for
   a. holding the cutting tool.
   b. providing power for rotation.
   c. transportation of the drill.
   d. finding the center of the hole.

2. Manufacturing is
   a. nonexistent in the United States.
   b. a synonym for machining.
   c. producing something.
   d. all about durable goods.

3. Equipment used to produce consumable goods is an example of
   a. society's dependence on machining.
   b. how manufacturing provides excellent careers.
   c. indirect machining.
   d. manufacturing in the United States.

4. Which manual machine tool has a CNC version?
   a. The lathe
   b. Sawing machine
   c. Grinder
   d. All of these choices

Name: ______________________________ Date: ______________

5. The drill press may be used when
   a. precise hole locations are needed.
   b. the part requires holemaking operations.
   c. the part requires abrasive machining.
   d. the sawing machine is not available.
6. The water jet machine uses a high-pressure stream of water containing
   a. chemicals.
   b. oils.
   c. abrasive grit.
   d. electric current.
7. A laser cuts material by using a concentrated
   a. beam of light.
   b. stream of water.
   c. electrical current.
   d. stream of air.
8. Precision machining can produce sizes within about 1/50th the thickness of a piece of paper, which is equal to
   a. .0010″
   b. .0001″
   c. .0100″
   d. .1000″
9. Extremely smooth, flat surfaces can be produced by a
   a. pedestal grinder.
   b. cylindrical grinder.
   c. bench grinder.
   d. surface grinder.

True or False

10. EDM machines are classified as manual machine tools. ________
11. Machining is connected to the manufacturing of durable goods. ________
12. Some machining operations can produce sizes to within .0001 inch or less of the desired size. ________
13. The United States is still a global leader in manufacturing. ________
14. Machine tools have little connection to everyday life. ________
15. Abrasive machining uses saw blades to perform cutting. ________
16. A milling machine can use either a vertical or horizontal spindle. ________
17. A lathe cuts a rotating piece of material to produce cylindrical parts. ________

Name: ______________________________ Date: ______________

Identification Matching

18. Match each industry to the machined parts it utilizes or produces.

a. Plastic	______	**1.** Suspension components, engine, and drivetrain
b. Medical	______	**2.** Forging dies
c. Automotive	______	**3.** Machined molds
d. Hand tool	______	**4.** Forming dies
e. Metal packaging	______	**5.** Knee and hip implants

19. Match the term with the description of its connection to the machining industry.

a. End products	______	**1.** Parts are often produced by tools called "molds" that are produced by machining
b. Manufacturing	______	**2.** Tools created by machining that are used to cut and/or form sheet metal to a desired shape
c. Plastic goods	______	**3.** Final manufactured items produced by machining that are used directly by consumers
d. Dies	______	**4.** Implants produced by machining for use in surgery
e. Medical devices	______	**5.** Producing an item

20. Select the appropriate machine tool process to produce the given product.

a. End products	______	**1.** Uses grinding wheels for either precision or non-precision machining
b. Manufacturing	______	**2.** Uses a highly concentrated beam of light
c. Plastic goods	______	**3.** Performs hole-making operations by feeding various types of rotating cutting tools into the material
d. Dies	______	**4.** Electrically charged wire or electrode used to machine material
e. Medical devices	______	**5.** Used to cut material to rough lengths or to remove large sections of material quickly
f. Laser machining	______	**6.** Uses abrasive grit in a very high-pressure water stream
g. Water jet machining	______	**7.** Used to produce cylindrical parts
h. Abrasive machining	______	**8.** Uses rotating cutters moved across a part to remove material

Short Answer

21. Briefly define machining in your own words.

____

____

22. Briefly describe how the machining industry is related to a common product you use every day.

____

____

23. Provide two examples of consumable items and explain how they indirectly rely on machining.

____

____

Name: ______________________ Date: ______________

24. Briefly describe the common features produced by both manual and CNC mills, as well as those unique to CNC milling.

____

____

25. Briefly define a machine tool.

____

____

Unit 2 Careers in Machining

Name: ______________________________ Date: ______________

Score: ______________ Text pages 16–23

Learning Objectives

After completing this unit, you should be able to:

- Identify careers in the machining industry
- Identify careers in fields related to machining
- Describe the job outlook in the machining field
- Describe effective job-seeking skills

Carefully read the unit, and then answer the following questions.

Multiple Choice

1. The most experienced employee in a conventional machine shop would likely be a
 a. machine operator.
 b. setup technician.
 c. toolmaker.
 d. CNC machinist.

2. A career field that requires troubleshooting and problem-solving skills as well as an understanding of the function and repair of mechanical, electrical, electronic, hydraulic, and pneumatic systems is a(n)
 a. setup technician.
 b. field service technician.
 c. industrial salesman.
 d. quality control inspector.

3. A career field title that focuses on the manufacturing process is a
 a. mechanical engineer.
 b. quality engineer.
 c. design engineer.
 d. manufacturing engineer.

4. Engineering technicians
 a. work with theory and design.
 b. work in the construction of machinery.
 c. never work with machinists.
 d. are focused on sales.

Name: ______________________________ Date: ______________

5. Entry-level employees who place parts in machine tools and run operations are called
   a. set-up technicians.
   b. operators.
   c. CNC machinists.
   d. programmers.
6. The science or practice of measurement is called
   a. metallurgy.
   b. inspection.
   c. quality control.
   d. metrology.
7. A very highly skilled CAM programmer concentrates on programming
   a. CNC milling machines.
   b. CNC lathes.
   c. CNC 5-axis milling machines.
   d. All of these choices are correct.

True or False

8. CNC operators are always highly skilled machinists. ________
9. Machining careers can provide advancement opportunities to management and engineering. ________
10. Today, the term machinist and mechanic are interchangeable. ________
11. The job outlook for machinists and toolmakers has been positive in recent years. ________
12. CNC operators are responsible for the quality of product, not the development of the CNC part program. ________

Identification Matching

13. Match the career job role to the appropriate job duty.

a.	Set-up technician	________	1. Uses CAD software to create models or engineering drawings
b.	Operator	________	2. Makes tooling to produce either plastic, glass, or die-cast metal parts
c.	Conventional machinist	________	3. Makes complex tools, jigs, and fixtures
d.	CNC machinist	________	4. Selects proper cutting tools and devices
e.	Programmer	________	5. Makes cutting tools consisting of punches and dies
f.	Toolmaker	________	6. Experienced in running almost every type of manual machine tool
g.	Die maker	________	7. Responsible for planning, scheduling, and purchasing
h.	Mold maker	________	8. Possesses the skills of both the setup technician and the operator
i.	Supervisor	________	9. Places parts in machines and continually runs a set operation
j.	Designer	________	10. Writes programs consisting of machine code

Name: ____________________ Date: ____________

Short Answer

14. How have occupations in the machining field changed over time?

____

____

15. Compare and contrast CNC and conventional machinists. How are their jobs similar? How are they different?

____

____

16. Briefly discuss the similarities and differences between die makers, mold makers, and toolmakers.

____

____

17. What type or types of machining careers might be of interest to you? Why? Which ones would not interest you? Why not?

____

____

18. Review the U.S. Department of Labor sponsored website (http://www.careeronestop.org/). List and describe two proven steps to find your next job.

____

____

Unit 3 Workplace Skills

Name: ______________________________ Date: ______________

Score: ______________ Text pages 24–38

Learning Objectives

After completing this unit, you should be able to:

- List personal skills needed for success in the machining field
- List technical skills needed for success in the machining field
- Explain the importance of the different personal skills needed in machining
- Explain the importance of different technical skills needed in machining
- Identify training opportunities and methods available to gain the skills required for the machining field
- Explain the reason for a career plan
- List key components of a resume
- Explain who could be used as references
- List key components of a cover letter
- Describe a portfolio and its importance
- Explain the purpose of a thank-you letter

Carefully read the unit, and then answer the following questions.

Multiple Choice

1. The ability to visualize relationships between moving parts is called
  a. mechanical aptitude.
  b. manual dexterity.
  c. eye-hand coordination.
  d. physical ability.

2. The ability to communicate effectively and to share information among peers is called
  a. persistence.
  b. interpersonal skills.
  c. public speaking.
  d. interpretation.

3. Companies may also need to provide training for specific or specialized areas and may send employees off-site or bring trainers on-site to meet that need; this kind of training is called
  a. RJT.
  b. JTO.
  c. OJT.
  d. JOT.

Name: ______________________________ Date: ______________

4. An evaluation/summary of what someone wants to accomplish during their working life is called a
   a. resume.
   b. portfolio.
   c. career plan.
   d. cover letter.
5. A person who is enrolled in a formal training program by an employer is a(n)
   a. journeyperson.
   b. apprentice.
   c. trainee.
   d. candidate.
6. After completing a company provided training program, a person receives the title of
   a. apprentice.
   b. mentor.
   c. journeyperson.
   d. trainer.
7. Working to recognize problems and how to solve them is called
   a. troubleshooting.
   b. critical thinking.
   c. technical skill.
   d. personal skill.
8. People who can provide information to an employer about your skills and attitude are
   a. mentors.
   b. trainers.
   c. references.
   d. acquaintances.

True or False

9. Instant gratification and completion is usually the norm for those in the machining field. ________
10. Even small lapses in attention can lead to large errors and huge losses of time and money. ________
11. Mechanical aptitude is the physical ability to precisely control hand movements. ________
12. All machining careers require a combination of mental and hands-on skills. ________

Short Answer

13. List and briefly describe the five major resume elements.

____

____

14. List five personal skills important in machining careers.

____

____

Name: ______________________________ Date: ______________

15. List the three parts of a cover letter.

____

____

16. Think about the personal workplace skills discussed in this unit related to you. Which skills do you feel you might already possess or are your current strengths? Where do you think you might need to improve?

____

____

17. Describe the importance of a career portfolio.

____

____

18. Explain the purpose of a thank-you letter.

____

____

19. List two technical skills required for a successful machining career and give one way each can be learned.

____

____

Unit 1 Introduction to Safety

Name: ______________________ Date: ______________

Score: ______________ Text pages 39–58

Learning Objectives

After completing this unit, you should be able to:

- Define OSHA
- Describe the role of OSHA
- Define NIOSH
- Describe the role of NIOSH
- Describe appropriate clothing for a machining environment
- Identify appropriate PPE used in a machining environment
- Describe proper housekeeping/behavior for a machining environment
- Describe the purpose of lockout/tagout (LOTO) procedures
- Define the term HCS
- Define the terms GHS, NFPA, and HMIS
- Identify GHS, NFPA, and HMIS labeling systems
- Define the term SDS
- Interpret SDS information
- Select the proper fire extinguisher application

Carefully read the unit, and then answer the following questions.

Multiple Choice

1. SDS stands for
  a. Safety Diagnostic Software.
  b. Safety Disclosure System.
  c. Sound Dampening System.
  d. Safety Data Sheet.

2. Disposal considerations are in the SDS section
  a. three.
  b. seven.
  c. nine.
  d. thirteen.

3. The term PPE refers to
  a. Personal Property Equipment.
  b. Personal Production Equipment.
  c. Personal Protective Equipment.
  d. Personal Product Equipment.

Name: ______________________________ Date: ______________

4. The most common personal safety equipment used in a machine shop is
   a. steel-toed shoes.
   b. safety glasses.
   c. long-sleeve shirt.
   d. shop apron.
5. The term *housekeeping* refers to
   a. the ability to clean the house.
   b. a janitor-type position.
   c. cleaning up only when major spills occur.
   d. keeping the work area clean and tidy.
6. The best method for removing metal chips from a machine is to use
   a. compressed air with a limited flow nozzle.
   b. your hands while wearing gloves.
   c. a brush or rag.
   d. a scraper and a pan.
7. If a small foreign particle enters the eye, what steps should be taken?
   a. Rub the eyelid to remove the particle.
   b. Gently pull the eyelid and flush with water.
   c. Remove the particle with a cotton swab.
   d. Sterilize the eye area with distilled water and remove with a cotton swab.
8. Before any repair or maintenance is performed on a machine, the OSHA standard of
   a. lockout/tagout should be followed.
   b. notifying the manufacturer should be followed.
   c. reporting of downtime should be followed.
   d. NFPA should be followed.
9. Companies rely on these to show details about the handling of hazardous materials.
   a. STEL
   b. NIOSH
   c. SDS
   d. HMIS
10. An HMIS rating of three (3) indicates
    a. minimal hazard.
    b. serious hazard.
    c. slight hazard.
    d. severe hazard.
11. NFPA stands for
    a. National Fire Protection Association
    b. National Fire Prevention Association
    c. National Fire Prevention Administration
    d. National Fire Protection Administration

Name: ______________________________ Date: ______________

12. HMIS stands for
  a. Hazardous Material Information System
  b. Hazardous Material Identification System
  c. Hazardous Material Identification Sheet
  d. Hazardous Material Information Sheet
13. The lowest temperature at which a substance can be ignited is its
  a. upper explosive limit.
  b. lower explosive limit.
  c. ignition level.
  d. flash point.
14. Images shown on GHS approved labels to visually show hazards are
  a. pictographs.
  b. photograms.
  c. pictograms.
  d. picturegrams.
15. The international system for hazardous material labeling is GHS, or
  a. Global Health Standard
  b. Globally Harmonized System
  c. Global Health System
  d. Globally Harmonized Standard
16. Safety glasses must have a rating of
  a. HCS Z87
  b. GHS Z87
  c. OSHA Z87
  d. ANSI Z87
17. HCS stands for
  a. Hazard Control Standard
  b. Hazard Communication Standard
  c. Hazardous Compliance Strategy
  d. Hazard Classification System

True or False

18. Different machining operations require different safety precautions. ________
19. Many accidents are caused by not paying attention and becoming distracted. ________
20. Safety is the responsibility of everyone. ________
21. You should always approach someone from the front if they are operating a machine. ________
22. OSHA requires some type of control or hearing protection if sound levels are above 115 dB (decibels) for 1/4 hour or over 90 dB for 8 hours. ________
23. For those who wear regular prescription eyeglasses, safety glasses are not required. ________
24. Compressed air can be used safely to remove metal chips, oils, solvents, or other chemicals from machinery. ________
25. Hazardous material information is contained in a document known as a Safety Data Sheet. ________

Name: ______________________________ Date: ______________

Identification Matching

26. Match the class of fire to the correct fire extinguisher.

a. Class A	________	**1.** Carbon dioxide
b. Class B	________	**2.** Water based
c. Class C	________	**3.** Dry chemical, powder, or foam

Short Answer

27. What does NIOSH stand for and what is NIOSH's purpose?

____

____

28. List three types of PPE.

____

____

29. What does *OSHA* stand for and what is OSHA's purpose?

____

____

30. What should anyone do with his or her jewelry when working around machinery?

____

____

31. List five good housekeeping rules.

____

____

32. Describe the OSHA requirement "lockout and tagout."

____

____

33. Describe how to interpret a rating of 4 in the red diamond in an NFPA hazard label.

____

____

34. List the three factors needed for fire to start.

____

____

Unit 2 Measurement Systems and Machine Tool Math Overview

Name: ______________________________ Date: ______________

Score: ______________ Text pages 59–80

Learning Objectives

After completing this unit, you should be able to:

- Differentiate between English and metric (SI) measurements
- Perform conversions between English and metric units
- Perform fractional and decimal math operations
- Perform conversions between fractions and decimals
- Solve formulas/equations using basic algebra
- Explain properties of basic geometry
- Determine complementary and supplementary angles
- Perform conversions between angular measurements in decimal degrees and degrees, minutes, and seconds
- Perform addition and subtraction of angular measurements
- Locate points in a Cartesian coordinate system
- Locate points in a polar coordinate system
- Utilize the Pythagorean theorem
- Solve right triangles using sine, cosine, and tangent trigonometric functions

Carefully read the unit, and then answer the following questions.

Multiple Choice

1. The order of operations when calculating a series of numbers is as follows:

a. Parentheses, Exponents, Addition, Division, Subtraction, Multiplication
b. Addition, Subtraction, Exponents, Multiplication, Division, Parentheses
c. Parentheses, Exponents, Multiplication, Division, Addition, Subtraction
d. Subtraction, Multiplication, Addition, Division, Exponents, Parentheses

2. How many millimeters are there in 1″?

a. 2.54
b. 0.254
c. 25.4
d. 254.4

3. The distance from one side of a circle to the other going through the center is called the

a. chord.
b. diameter.
c. radius.
d. segment.

Name: ______________________________ Date: ______________

4. How many degrees are there in a circle?
   a. 180 degrees
   b. 90 degrees
   c. 270 degrees
   d. 360 degrees
5. A complementary angle is calculated by subtracting the known angle from
   a. 180 degrees.
   b. 90 degrees.
   c. 270 degrees.
   d. 60 degrees.
6. The Cartesian coordinate systems use two- and three-dimensional location axes, which are divided into __________ quadrants.
   a. one
   b. two
   c. three
   d. four
7. In right triangle calculations using trigonometry, the longest side of a triangle is called the
   a. adjacent side.
   b. opposite side.
   c. hypotenuse.
   d. 90-degree angle.
8. Sine is a ratio between the lengths of the
   a. opposite side and hypotenuse.
   b. hypotenuse and angle.
   c. adjacent side and opposite side.
   d. adjacent and hypotenuse.
9. What is the inch equivalent of 1 mm?
   a. .03937
   b. .3937
   c. 3.937
   d. 39.37
10. The English measurement system is based on
   a. thousandths.
   b. inches.
   c. millimeters.
   d. feet.
   e. meters.
11. A line, circle, or arc that touches a circle or an arc at only one point is called
   a. tangent.
   b. chord.
   c. angular.
   d. parallel.

Name: ______________________________ Date: ______________

12. Locations given by an angle and a distance from an origin are
   a. Cartesian coordinates.
   b. polar coordinates.
   c. X, Y coordinates.
   d. angular coordinates.
13. The Pythagorean Theorem is
   a. $a^2 + b^2 = c^2$
   b. $a^2 - b^2 = c^2$
   c. $a + b = c^2$
   d. $a + b = c$

True or False

14. The complementary angle of 70 degrees is 20 degrees. ________
15. The metric system commonly uses fractional units, while the English system favors decimals. ________
16. Points in the 2D Cartesian coordinate system are always recorded in Y, X format. ________
17. Order of operations is the principle that certain mathematical operations must be done in a certain order to correctly perform calculations. ________
18. The circumference of a circle is the distance across the center of the circle. ________
19. A circle contains 360 degrees. ________
20. The sum of the angles of every triangle is always 180 degrees. ________
21. Parallel lines or surfaces are 90 degrees to each other. ________

Computation

22. Place in order the following fractions from smallest to largest: 15/16, 3/8, 7/32, 27/64, 21/32, 1/4.

____

____

23. What are the decimal equivalents of the fractions from the previous question? (Write answers to the nearest 0.0001″.)

____

____

24. What is 3/16 + 9/32?

____

____

25. What is 5/8 + 9/16?

____

____

26. What is 7/16 + 0.005?

____

____

Name: ______________________________ Date: ______________

27. What is 1/2 − 1/8?

28. What is 1 1/4 − 5/8?

29. What is 3/8 − 0.002?

30. What is 1/2 × 9/16?

31. What is 3/4 × 1 3/8?

32. What is 200 ÷ 1/4?

33. What is 1/2 × 1.505?

34. What is (0.75 × 116.50 − (13 − 8) × 2) / (7 × 8 − 4)?

35. What is the sum of the complementary angle of 55 degrees and the supplementary angle of 15 degrees?

36. How many inches are there in a millimeter?

37. What is the metric equivalent of 0.575″ to the nearest 0.01 mm?

38. Convert 5.635″ to mm.

39. Convert 29.27 mm to inches.

Name: ______________________ Date: ______________

40. Solve the formula $A = B/C$ for A when $B = 130$ and $C = 5/16$.

41. Solve the formula $A = (B - C)/D$ for B when $A = 1/4$, $C = 1.25$, and $D = 4.500$.

42. Convert 60.46 degrees to degrees, minutes, and seconds. Round to the nearest second.

43. What is the circumference of a circle that is 5″ in diameter?

44. Find the length of side "a" of a right triangle to the nearest 0.001″ if side "b" = 2.545″ and "c" = 5.850″.

45. Find the length of side "b" of a right triangle to the nearest 0.01 mm if side "a" = 6.958″ and "c" = 9.250″.

46. Use sine to find the angle in a right triangle if the opposite side = 2.500″ and the hypotenuse = 6.500″?

Short Answer

47. List the order of operations.

48. What is the Cartesian coordinate system?

49. List the three trigonometric functions as the ratios of the sides of a right triangle.

50. Define the terms *radius* and *diameter* of a circle.

Unit 3 Semi-Precision Measurement

Name: ______________________ Date: ______________

Score: ______________ Text pages 81–95

Learning Objectives

After completing this unit, you should be able to:

- Define semi-precision measurement
- Define comparative measurement
- Explain basic care of common semi-precision measuring instruments
- Read an English rule to within 1/64 of an inch
- Read an English (decimal) rule to within 1/100 of an inch
- Read a metric rule within 0.5 mm
- Explain the uses of semi-precision calipers
- Explain the uses of squares
- Explain the uses of the combination set
- Explain the uses of protractors
- Read protractors within 1 degree
- Explain the uses of common semi-precision fixed gages

Carefully read the unit, and then answer the following questions.

Multiple Choice

1. The term *semi-precision measurement* usually refers to measurements when tolerances are within
  a. 1/64″.
  b. 1/100″.
  c. 0.5 mm.
  d. All of these choices are correct.

2. A common measuring tool for measuring linear distances in the machining trade is a
  a. tape measure.
  b. steel rule.
  c. yard rule.
  d. ruler.

3. Outside and inside spring calipers are considered what type of measuring tools?
  a. Direct measuring
  b. Comparison measuring
  c. Indirect measuring
  d. Transfer measuring

Name: ______________________________ Date: ______________

4. Adjustable squares can be used to check
   a. perpendicularity.
   b. roundness.
   c. flatness.
   d. All of these choices are correct.
5. The best tool to check for an inside radius contour is a(n)
   a. angle gage.
   b. pitch gage.
   c. radius gage.
   d. ball gage.
6. The allowable variation from a print dimension is the
   a. tolerance.
   b. allowance.
   c. error.
   d. deviation.
7. Examine the part and metric rule shown in Figure 2.3.4 of the textbook. The linear distance, to the nearest 0.5 mm, is
   a. 45 mm.
   b. 35 mm.
   c. 30.5 mm.
   d. 40 mm.
8. A combination square can measure angles that are
   a. 30 and 60 degrees.
   b. 30 and 90 degrees.
   c. 45 and 90 degrees.
   d. 60 and 90 degrees.
9. The smallest graduation on a metric rule is usually
   a. .1 mm.
   b. .5 mm.
   c. .05 mm.
   d. 5 mm.

True or False

10. Tolerances determine the type of measuring tool to be used for inspection. ________
11. Comparative measurements do not involve actual numerical values. ________
12. Examine the rule shown in Figure 2.3.8 (B) of the textbook. An 8.453″ measurement is taken using the 1/64″ graduation. ________
13. Examine the rule shown in Figure 2.3.9 (A) of the textbook. An 8.800″ measurement is taken using the 0.010″ graduation. ________
14. A combination square can be used to measure a 30-degree angle. ________
15. Checking for square means checking parallelism. ________

Name: ____________________ Date: ____________

Identification Matching

16. Select the best semi-precision tool to measure the listed dimension.

a. Screw pitch gage	______	1. Make a comparative measurement within 10 degrees of perpendicular
b. Combination square	______	2. Measure the inside diameter of a pipe
c. Outside spring caliper	______	3. Check the size of a corner radius
d. Inside spring caliper	______	4. Check a slot depth
e. Die maker's square	______	5. Make a comparative check of a 45-degree angle
f. Protractor	______	6. Determine the distance between threads
g. Adjustable square	______	7. Measure a linear distance
h. Radius gage	______	8. Compare one unknown angle to another
i. Steel rule	______	9. Measure a 15-degree angle
j. Bevel	______	10. Measure the outside diameter of a cylindrical part

Short Answer

17. Briefly explain the use of a screw pitch gage.

18. Briefly explain how to measure an outside diameter with a spring caliper.

19. List all the parts of a combination set.

20. Briefly explain basic care of semi-precision measuring instruments.

21. Examine the part and protractor shown in Figure 2.3.26 of the textbook. What is the angle, within 1 degree, between the two part surfaces?

Unit 4 Precision Measurement

Name: ______________________ Date: ______________

Score: ______________ Text pages 96–154

Learning Objectives

After completing this unit, you should be able to:

- Explain basic care of precision measuring tools
- Explain the purpose of calibration
- Explain the use of common precision fixed gages
- Describe the purpose of a surface plate
- Explain the principle of how micrometers work
- Describe the process of micrometer calibration
- Name different types of micrometers
- Read various types of micrometers
- Read various types of vernier scale tools
- Explain uses of precision transfer-type measuring instruments
- Explain uses of indicators
- Explain uses of gage blocks
- Explain the uses of sine tools
- Explain methods for measuring surface finishes
- Explain the purpose of a toolmaker's microscope
- Explain the purpose of an optical comparator
- Describe the basic principle of a coordinate measuring machine (CMM)

Carefully read the unit, and then answer the following questions.

Multiple Choice

1. Precision measurement should be used when tolerances are ______________ or smaller.
   a. 0.020″
   b. 0.005″
   c. 0.100″
   d. All of these choices

2. When determining a gage class of tolerance and accuracy, a series of numbers is given to rank the accuracy. What is the most accurate gage class listed below?
   a. Class Z
   b. Class X
   c. Class XXX
   d. Class ZZ

Name: ______________________________ Date: ______________

3. Which gage is the best for checking unthreaded external diameters (OD)?
   a. Taper plug gage
   b. Ring gage
   c. Gage blocks
   d. Threaded snap gage
4. The best precision tool to check the perpendicularity of two part surfaces is a
   a. combination square.
   b. surface plate.
   c. beveled solid square.
   d. thick steel rule.
5. Which of the following is a type of micrometer?
   a. Multiple Anvil
   b. Outside
   c. Screw Thread
   d. All of these choices
6. On tools with a 50-part vernier scale, each inch on the main scale is broken into ______________ parts measuring 0.050″ each.
   a. 25
   b. 50
   c. 20
   d. 30
7. The best way to calibrate a standard 1″–2″ micrometer is to use a(n)
   a. inside micrometer.
   b. digital caliper.
   c. adjustable parallel.
   d. gage block.
8. The angle of a dial indicator's contact point to the workpiece is very important during setup to prevent
   a. damage to the workpiece.
   b. damage to the indicator holder.
   c. errors in indicator readings.
   d. All of these choices are correct.
   e. None of these choices is correct.
9. The center-to-center distance of the sine bar roll is representative of which side of a right triangle?
   a. Hypotenuse
   b. Opposite side
   c. Adjacent side
   d. Sine angle
10. The most accurate method to measure Ra surface roughness is with a(n)
   a. comparator gage.
   b. profilometer.
   c. optical comparator.
   d. surface plate.

Name: ______________________ Date: ______________

11. Verifying the accuracy of a measuring tool is called
    a. calibration.
    b. confirmation.
    c. inspection.
    d. authentication.
12. An inch-based micrometer with a vernier scale can measure as accurately as
    a. .010″.
    b. .001″.
    c. .0001″.
    d. .00001″.
13. Measuring small parts and dimensions by projecting a magnified image on a screen is performed on a(n)
    a. optical comparator.
    b. optical calibrator.
    c. surface comparator.
    d. surface calibrator.
14. The diameter of a .755″ hole can best be measured using a micrometer and a(n)
    a. small hole gage.
    b. telescoping gage.
    c. adjustable parallel.
    d. inside spring caliper.

True or False

15. Dropping precision tools will not affect their accuracy. ________
16. A surface plate is used as an accurate reference surface with other precision tools to aid with some measurement tasks. ________
17. All vernier measuring tools have two scales on them. ________
18. A toolmaker's microscope is used to inspect very small parts. ________
19. A dial indicator looks and works much like a car's speedometer. ________
20. Straight edges are never used together with feeler gages. ________

Name: ______________________________ Date: ______________

Identification Matching

21. Select the best measuring tool for the given application.

a. Outside micrometer	________	**1.**	Determines flatness of a part
b. Vernier caliper	________	**2.**	Provides a reference plane for using other tools
c. Telescoping gage	________	**3.**	Measures an inside diameter of 1.75″ 1/−0.001″
d. Straight edge	________	**4.**	Measures a round bar 0.500″ 1 0.001/−0.002″
e. Pin gage	________	**5.**	Determine the size of hole
f. Surface plate	________	**6.**	Measures 100 pieces of 2.500″ square blocks
g. CMM dimensions	________	**7.**	Measures a large casting with numerous critical dimensions
h. Toolmaker microscope	________	**8.**	Measures an 8″ long bar within 0.005″
i. Optical comparator	________	**9.**	Check the profile/shape of a thread
j. Plunge indicator	________	**10.**	Inspect a part 0.05″ × 0.25″ × 0.020″ 1/−0.001″

Computations

22. If you were to select pin gages to check a hole with a 0.625″ dia. +0.002″/−0.003″, what pin sizes should be used for the go and no-go gages? Provide a three-decimal answer.

____

____

23. Determine the best gage block buildup for a 36-degree angle using a 10″ sine bar. Refer to pg. 105, Figure 2.4.25 in the textbook as reference for block sizes in the 88-piece set.

____

____

24. Calculate the angle if the gage block build is 2.3568″ on a 10″ sine bar.

____

____

Name: ______________________ Date: ______________

25. Record the following vernier scale readings.

a. ______________

b. ______________

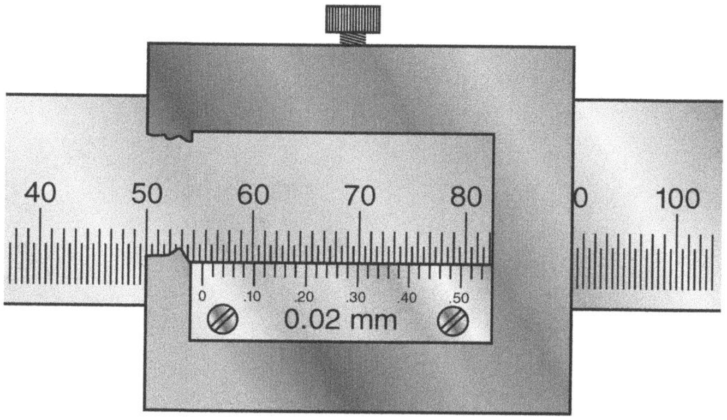

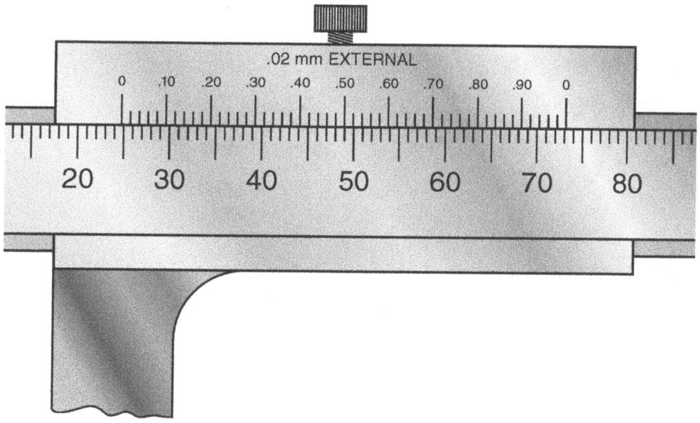

Name: ______________________________ Date: ______________

26. Record the following micrometer readings.

a. ______________

b. ______________

c. ______________

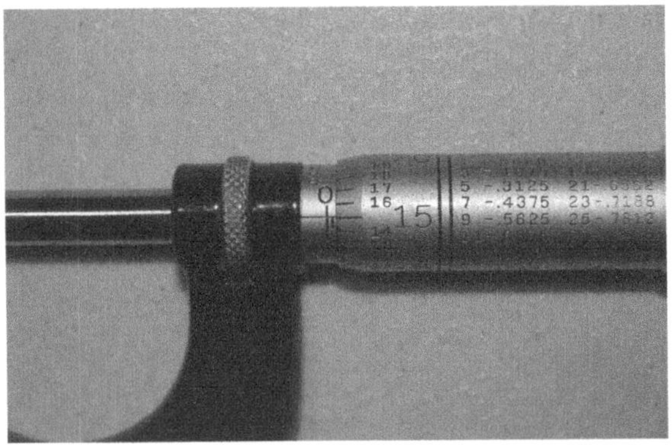

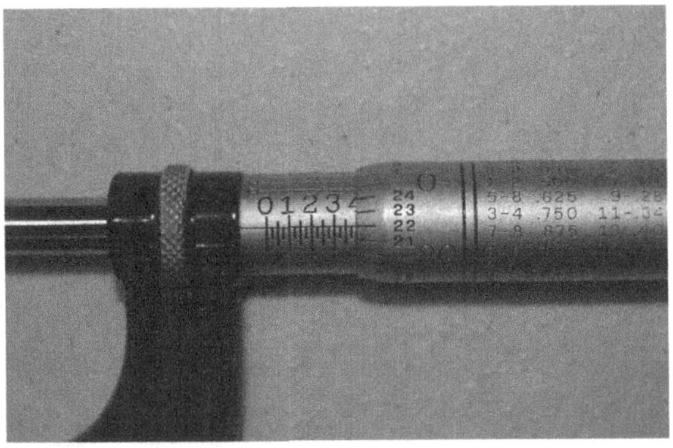

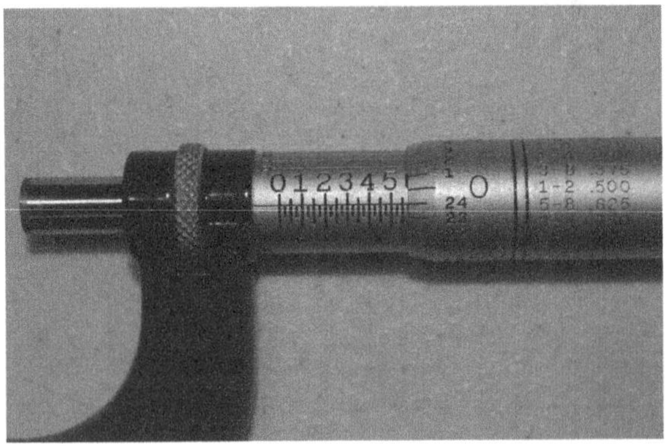

27. Record the following depth micrometer readings.

a. ______________

b. ______________

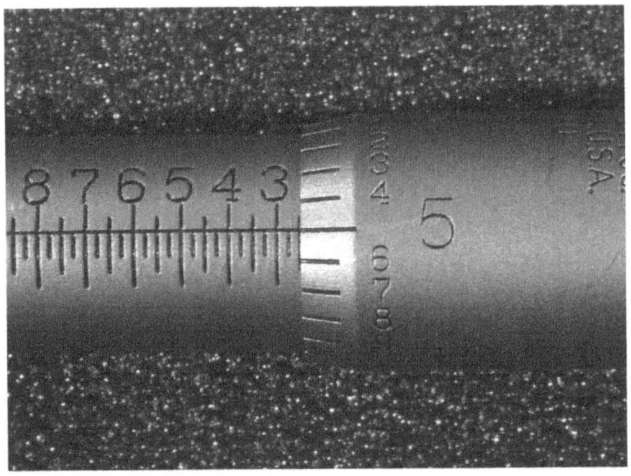

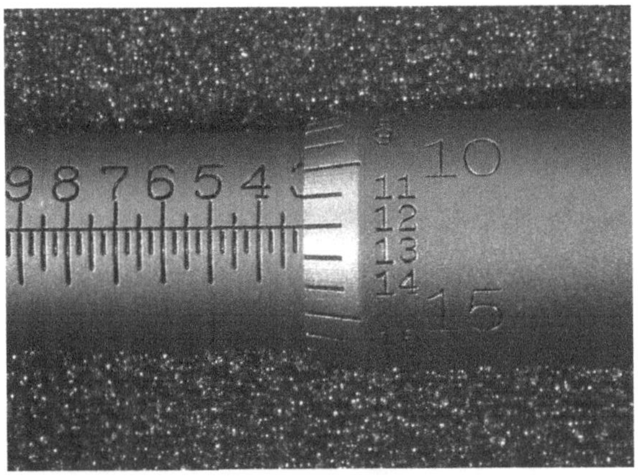

Name: ______________________________ Date: ______________

Short Answer

28. Describe the key components and explain the operation of a coordinate measuring machine (CMM).

____

____

29. Describe the process of wringing gage blocks.

____

____

30. Explain how to use a go and no-go ring gage to check an external thread.

____

____

31. List three transfer measuring tools.

____

____

32. Explain the process of calibrating a 0″–1″ micrometer.

____

____

33. Briefly explain the operating principle of a micrometer.

____

____

Unit 5 Quality Assurance, Process Planning, and Quality Control

Name: ______________________ Date: ______________

Score: ______________ Text pages 155–164

Learning Objectives

After completing this unit, you should be able to:

- Define quality assurance
- List the major parts of a process plan
- Define quality control
- Explain the purpose of quality control
- Describe a sampling plan
- Explain the purpose of an inspection plan
- List the major elements of an inspection plan
- Define statistical process control (SPC)
- Explain the purpose of SPC
- Explain the purpose of an X-bar chart
- Explain the purpose of an R-chart

Carefully read the unit, and then answer the following questions.

Multiple Choice

1. Which of the following contains information such as required machines and tools, material selection, and machine settings?
  a. sample plan.
  b. inspection plan.
  c. quality plan.
  d. process plan.

2. A machinist may be called upon to collect dimensional data during a production. The best way to do this is to use a(n)
  a. control plan.
  b. control chart.
  c. production chart.
  d. inspection chart.

3. An R-chart or range chart is used to determine the
  a. variation of each sample.
  b. actual size of the sample.
  c. tool wear.
  d. machine stress during a cut.

Name: ______________________________ Date: ______________

4. During a production run, the operator notices that the R-chart is continually increasing over five subgroups but is still under the UCL. What should the operator do?
   a. Call the supervisor.
   b. Keep watching the graph and notice differences.
   c. Shut down the machine.
   d. Adjust the cut to make the increasing line decrease.
5. The UCL and LCL are determined by the part
   a. size.
   b. material.
   c. tolerance.
   d. process plan.
6. Quality control follows a plan of how often to inspect parts and how many to inspect. These steps are found on a(n)
   a. inspection plan.
   b. control plan.
   c. sampling plan.
   d. production plan.
7. A document that lists dimensions to check and what tools to use is a(n)
   a. process plan.
   b. inspection plan.
   c. sampling plan.
   d. production plan.
8. The selection of the measuring tool to be used is based on the
   a. tolerance.
   b. inspection plan.
   c. basic size.
   d. process plan.
9. SPC stands for
   a. sampling program control.
   b. sampling process control.
   c. statistical program control.
   d. statistical process control.
10. An X-bar chart graphs the
   a. mean.
   b. range.
   c. median.
   d. tolerance.

Name: ______________________________ Date: ______________

True or False

11. Quality assurance is the plan that guides the actions and performance of the machining process. ______
12. Sampling plans tell how often and how many parts to inspect. ______
13. The purpose of statistical process control (SPC) is to predict the consistency of an operation and guide adjustments needed to maintain process stability. ______

Identification Matching

14. Match the appropriate term to the defined action.

a. Sampling plan	______	**1.** Shows trends of average size of a part
b. Inspection plan	______	**2.** Explains what measuring tools to use
c. Process plan	______	**3.** Uses smallest and largest dimensions to calculate variance
d. X-bar chart	______	**4.** Defines the steps needed to make a part
e. R-chart	______	**5.** Determines the number of parts to inspect

Computation

15. What is the average of a subgroup with part sizes of 0.640, 0.635, 0.642, 0.637, 0.641, 0.643, and 0.639?

____

____

16. Using the decimal values given in the previous question, what is the range of the subgroup?

____

____

Short Answer

17. An R-chart line is continually increasing. What is happening to this operation?

____

____

18. An R-chart line is continually decreasing. What does this tell about the operation?

____

____

19. An X-bar chart is showing a particular machined size is starting to move toward the LCL. Briefly explain what is happening and what steps should be taken.

____

____

Name: ______________________ Date: ______________

20. What is an inspection plan, and what is its purpose?

____

____

21. Explain the term *critical dimension* and how it affects the inspection planning stage.

____

____

22. What is Quality Assurance (QA) and Quality Control (QC), and how do they differ?

____

____

Unit 6 Metal Composition and Classification

Name: ______________________ Date: ______________

Score: ______________ Text pages 165–182

Learning Objectives

After completing this unit, you should be able to:

- Describe the difference between ferrous and nonferrous metals
- Compare and contrast low-, medium-, and high-carbon steels
- Define an alloy
- List common alloying elements
- Describe the differences/similarities between steel and cast iron
- Identify compositions of AISI/SAE steel designations
- Identify compositions of UNS carbon and alloy steels designations
- Identify compositions of AA/IADS aluminum alloy designations
- Identify UNS designations for stainless steels
- Identify UNS designations for cast iron
- Identify UNS designations for nonferrous alloys

Carefully read the unit, and then answer the following questions.

Multiple Choice

1. All ferrous metals contain ______________.

a. steel
b. iron
c. oxygen
d. zinc

2. Steels that have additional elements added to the iron and carbon are called ______________ steels.

a. wrought
b. alloy
c. pure
d. carbon

3. When carbon rises to 1.7%–4.5%, the material is known as

a. carbon steel.
b. tool steel.
c. cast iron.
d. wrought iron.

Name: ____________________ Date: ____________

4. What is the percentage of carbon content of the following material: AISI 4135?
   a. .41
   b. .35
   c. .13
   d. .45
5. Which of the stainless steels below are not magnetic?
   a. Austenitic
   b. Ferrite
   c. Martensitic
   d. None of these choices
6. Brass is an alloy of
   a. copper and zinc.
   b. copper and tin.
   c. copper and aluminum.
   d. None of these choices is correct.
7. Bronze is an alloy of
   a. copper and zinc.
   b. copper and tin.
   c. copper and aluminum.
   d. None of these choices is correct.
8. Superalloys are ____________________ based metals.
   a. chromium
   b. titanium
   c. nickel
   d. tungsten
9. The UNS prefix for cast irons is
   a. F.
   b. C.
   c. I.
   d. J.
10. 7075 aluminum is an aluminum alloy containing
   a. copper.
   b. silicon.
   c. manganese.
   d. zinc.

True or False

11. Cast iron is very hard and brittle. ________
12. Ferrous and nonferrous metals contain iron. ________
13. All of the metal numbering systems use similar methods for their classification. ________
14. Malleable and ductile cast irons have the ability to flex and stretch before breaking. ________
15. 1050 aluminum is 99.50% pure. ________

Name: ______________________________ Date: ______________

Identification Matching

16. Select the best metal for the part or tool listed.

a. Aluminum	______	**1.** Car suspension part
b. Cast iron	______	**2.** Handrails
c. Tool steel	______	**3.** Jet engine part
d. Cold-rolled steel (CRS)	______	**4.** Food conveyor part
e. Martensitic stainless steels	______	**5.** Aircraft support brace
f. Austenitic stainless steels	______	**6.** Axle bushing
g. Wrought iron	______	**7.** Knee/hip joint
h. Bronze	______	**8.** Knife blade
i. Titanium	______	**9.** Punch and die
j. Superalloy	______	**10.** Base for a machine

Short Answer

17. What is tool steel? How is it the same as cast iron? How is it different?

____

____

18. Compare and contrast low-, medium-, and high-carbon steels.

____

____

19. What is the carbon content of 7240 steel?

____

____

20. What does a "B" in an alloy steel designate?

____

____

21. The UNS prefix "S" classifies what type of metals?

____

____

22. What UNS prefix classifies titanium alloys?

____

____

23. The UNS prefix "C" classifies what type of metals?

____

____

24. What UNS prefixes classify superalloys?

____

____

Unit 7 Heat Treatment of Metals

Name: ______________________________ Date: ______________

Score: ______________ Text pages 183–199

Learning Objectives

After completing this unit, you should be able to:

- Describe common heat-treatment processes
- Describe different types of heat-treating equipment
- Describe safety procedures and PPE for heat treating
- Explain Rockwell and Brinell hardness scales
- Compare and contrast Rockwell and Brinell hardness testing methods

Carefully read the unit, and then answer the following questions.

Multiple Choice

1. Direct hardening of steel can only be performed if the
  a. part contains less than .30% carbon.
  b. part contains more than .30% hydrogen.
  c. part contains more than .30% silicon.
  d. part contains more than .30% carbon.

2. Induction hardening uses
  a. cyanide.
  b. nitrogen.
  c. electrical current.
  d. chemicals.

3. The process that returns metal to its original hardness is
  a. annealing.
  b. normalizing.
  c. tempering.
  d. carburizing.

4. If a machinist wanted to reduce stress in a casting after machining operations are complete, what heat treatment would they use?
  a. Annealing
  b. Normalizing
  c. Tempering
  d. Nitriding

Name: ______________________________ Date: ______________

5. The diamond Brale penetrator and 60 kg preload are used on what hardness tester?
   a. Brinell
   b. Shore
   c. Rockwell
   d. SPIC
6. To relieve stress and provide better hardening results on forged parts, first perform
   a. annealing.
   b. normalizing.
   c. tempering.
   d. carburizing.
7. The cooling process after heating is called
   a. immersing.
   b. submerging.
   c. plunging.
   d. quenching.
8. What hardness testing scale uses a 1/16″ diameter steel ball and 100 kg major load?
   a. Rockwell C
   b. Brinell
   c. Rockwell B
   d. Rockwell A
9. Raising the elevating screw on a Rockwell hardness tester
   a. applies the minor load.
   b. applies the major load.
   c. releases the minor load.
   d. releases the major load.
10. A Rockwell C scale test is performed using the Brale diamond and a major load of
   a. 60 kg.
   b. 75 kg.
   c. 100 kg.
   d. 150 kg.

True or False

11. Treat every piece of metal as hot when working in a heat-treat area. ________
12. The Rockwell C scale can be used to test hardened tool steel. ________
13. The diameter of a ball impression is used when determining hardness using the Brinell Hardness System. ________
14. Annealing involves a quick cooling process to soften the material. ________
15. All furnaces perform the same basic task of heating metal to the required temperatures. ________

Name: ______________________ Date: ______________

Identification Matching

16. Select the appropriate heat-treatment process for the given process.

a. Case hardening ________

b. Surface hardening ________

c. Nitriding ________

d. Tempering ________

e. Annealing ________

f. Normalizing ________

g. Carburizing ________

1. Dipping or rolling in a carbon-rich powdered material to add an outer layer of carbon to steel

2. Returns metals to their original pre-hardened condition

3. Removes stresses and makes the structure more consistent

4. Creates a thin, hard outer shell on low carbon steels

5. Sealed furnace containing nitrogen gas to produce a hard-outer shell

6. Performed after hardening to decrease the steel's hardness, increase toughness

7. Only an outer layer is hardened, center remains soft

Short Answer

17. Briefly describe an atmospheric furnace and its benefit.

____

____

18. Which hardness scale uses a 10-mm carbide ball to make an indentation on a test sample?

____

____

19. Briefly describe the two most common heat-treatment processes typically applied to aluminum.

____

____

20. List four commonly used quenching media.

____

____

21. What hardness testing scale uses a Brale diamond and a 150 kg major load?

____

____

22. What process is used on aluminum to increase hardness?

____

____

23. List three PPE items (in addition to standard items) to use when performing a heat-treatment process.

____

____

Unit 8 Maintenance, Lubrication, and Cutting Fluid Overview

Name: ____________________ Date: ____________

Score: ____________ Text pages 200–210

Learning Objectives

After completing this unit, you should be able to:

- Explain the importance of a routine maintenance program
- List different methods of machine tool lubrication
- List routine machine tool maintenance inspection points
- Explain the purposes of cutting fluids
- Contrast the two basic types of cutting fluids
- Explain the methods of applying cutting fluids
- Explain basic safety precautions when using cutting fluids

Carefully read the unit, and then answer the following questions.

Multiple Choice

1. Before using any liquid solvent, grease, or oil, you should
  a. read the machine manual to check recommendations.
  b. read the PM shop report.
  c. check machine levels.
  d. read the SDS and PPE requirements.

2. Lubricants are used for
  a. cooling parts.
  b. minimizing friction.
  c. preventing seizures.
  d. All of these choices are correct.

3. Water-based cutting fluids are called
  a. soluble fluids.
  b. straight oils.
  c. chemical-based.
  d. water fluids.

4. Cutting fluids that contain no oil are called
  a. soluble oils.
  b. straight oils.
  c. synthetics.
  d. chemical-based.

Name: ______________________________ Date: ______________

5. When using soluble oils and synthetic cutting fluids, it is important to check the concentration levels of the mixture. A handheld device used for this is called a
   a. fraction meter.
   b. ratio meter.
   c. refractometer.
   d. soluble meter.
6. When a machine slide becomes loose, it can be tightened by adjusting the
   a. gib.
   b. wedge.
   c. taper.
   d. screw.
7. A fitting used to apply grease is called a
   a. ball oiler
   b. zerk
   c. reservoir
   d. cup
8. An additive that helps oil stick to cutting tools and workpiece is a(n)
   a. wetting agent.
   b. extreme pressure additive.
   c. sulfur.
   d. chlorine.
9. Cutting fluid that is held in tank and recirculated is part of a ______________ system.
   a. MQL
   b. mist
   c. emulsifiable
   d. flood
10. The ability of a cutting fluid to reduce friction is called
   a. lubrability.
   b. lubrapacity.
   c. lubricity.
   d. is not a real word but something the other authors listed as a choice.

True or False

11. Preventive maintenance tasks usually need to be done at specific intervals. ________
12. Any oil or grease will work to lubricate a machine tool. ________
13. Chemical-based fluids that are combined with water should never be used for magnesium. ________
14. Water-soluble fluids provide the best cooling level for machining. ________
15. A refractometer can be used to determine concentration levels of mixed fluids. ________

Name: ______________________ Date: ______________

Short Answer

16. List three reasons why cutting fluids are used.

17. List five nonliquid cutting compounds.

18. What are two major concerns when using synthetic oils?

19. What is MQL, and how does it benefit tool lubrication?

20. Explain the term backlash.

21. What are the two main types of cutting fluids, and how do they differ?

22. List routine inspection points for machine tool maintenance.

Unit 1 Understanding Drawings

Name: ______________________ Date: ______________

Score: ______________ Text pages 211–246

Learning Objectives

After completing this unit, you should be able to:

- Explain title block information
- Identify line types
- Identify orthographic views used on engineering drawings
- Identify isometric views
- Explain basic symbols and notation used on engineering drawings
- Define tolerance
- Explain unilateral, bilateral, and limit tolerances
- Determine allowances and classes of fit for cylindrical components
- Define geometric dimensioning and tolerancing (GD&T) and explain its purpose
- Identify basic GD&T symbols
- Explain the elements of basic GD&T feature control frames

Carefully read the unit, and then answer the following questions.

Multiple Choice

1. If a machinist needed to determine the revision of the print, where should he or she look on the print?
   - **a.** Body of the drawing
   - **b.** BOM list
   - **c.** Title block
   - **d.** Above the title block
2. The method of representing a three-dimensional object in a two-dimensional format is called
   - **a.** isometric projection.
   - **b.** orthographic projection.
   - **c.** single-view projection.
   - **d.** three-view projection.
3. When creating a drawing that has alternating or rotating parts, what line would be best to illustrate alternate positions?
   - **a.** Hidden
   - **b.** Phantom
   - **c.** Center
   - **d.** Solid

Name: ______________________________ Date: ______________

4. In order to show the internal features of a part, a __________ view can be used.
- **a.** front
- **b.** side
- **c.** section
- **d.** rotated

5. Convert the following unilateral tolerance to a bilateral tolerance: 2.500″ + 0.006″/−0.000″.
- **a.** 2.505″ + 0.001/−0.005″
- **b.** 2.504″ + 0.002/−0.004″
- **c.** 2.503″ + 0.003/−0.003″
- **d.** All of these choices
- **e.** None of these choices

6. The relationship between the sizes of two mating parts is called the
- **a.** class of fit.
- **b.** allowance.
- **c.** tolerance.
- **d.** interference.

7. Circularity applies to a single cross section of a feature, while cylindricity applies
- **a.** to the entire length of the feature.
- **b.** only to the LMC.
- **c.** the diameter of the feature.
- **d.** only to the MMC.

8. The term *MMC* refers to
- **a.** minimum material condition.
- **b.** minimum machining condition.
- **c.** maximum material condition.
- **d.** maximum machining condition.

9. The name of the three-dimensional view that is sometimes shown on a print to help visualize an object is
- **a.** orthographic view
- **b.** isometric view
- **c.** dimensional view
- **d.** 3-D view

10. The rectangular box used in GD&T to show the tolerance of a part characteristic is a
- **a.** tolerance control frame.
- **b.** dimension control frame.
- **c.** feature control frame.
- **d.** geometric control frame.

11. Tolerances can add up when finding an unknown or reference dimension. This is called a(n)
- **a.** accumulated tolerance.
- **b.** stacked tolerance.
- **c.** accrued tolerance.
- **d.** additional tolerance.

Name: ________________________________ Date: ______________

12. A dimension of 1.000 + .002/−.001 is an example of a(n)
   a. unilateral tolerance.
   b. limit tolerance.
   c. geometric tolerance.
   d. bilateral tolerance.

True or False

13. Radii on the inside and outside corners of a workpiece are sometimes called fillets and rounds. ________
14. General notes can be used to establish tolerances. ________
15. A basic size includes tolerances of the part. ________
16. A bonus tolerance can be given for position dimensions that are in the LMC state. ________
17. If a machine operator has an issue with a dimension on the print, he or she can measure the print to get the correct answer. ________
18. Two surfaces can be flat but not parallel. ________
19. The depth of a slot is a feature of size. ________

Identification Matching

20. Match the type of line to its description.

a. Object	________	1. A thin, angled line with an arrowhead on one end that points to a specific feature or detail
b. Hidden	________	2. A thin line that has arrowheads at the ends and shows a basic size
c. Center	________	3. A line that shows alternate positions of a part or outlines of adjacent parts
d. Dimension	________	4. A thick and continuous line showing part edges
e. Leader	________	5. A thick line that is drawn as one long and two short dashes alternately spaced, used to create a section
f. Extension	________	6. A thin dashed line showing edges that cannot be seen
g. Phantom	________	7. A line that extends from the edges of an object or feature is used to show a dimension
h. Cutting plane	________	8. A thin line broken into alternating long and short dashes that can show symmetry

Computations

21. Determine the total tolerance, upper limit, and lower limit for the following dimensions:

a. $\frac{1}{2} \pm \frac{1}{64}$ ________

b. $\frac{3}{4} + \frac{1}{64}/-0$ ________

c. $\frac{5}{8} + 0/\frac{1}{64}$ ________

d. 1.000 ± 0.003 ________

e. 1.0625 + 0/−0.0005 ________

f. 0.625 + 0.001/−0 ________

Name: ______________________ Date: ______________

22. What is the MMC of a shaft with a dimension of 1.250 + 0/−0.001?

23. What is the LMC of a shaft with a dimension of 1.750 + 0/−0.002?

24. What is the LMC of a hole with a dimension of 0.375 ± 0.001?

25. What is the MMC of a hole with a dimension of 0.875 ± 0.002?

26. Rewrite the limit tolerance as a basic size with an equal bilateral tolerance: 2.515/2.505

27. If a shaft is dimensioned at $\frac{1.0021}{1.0018}$ and the hole of the mating part is dimensioned at $\frac{1.0025}{1.0031}$, is the allowance positive or negative? What is the amount of the allowance?

Short Answer

28. If the scale on an engineering drawing is 1:2, is the drawing larger or smaller than the actual object?

29. Briefly define geometric dimensioning and tolerancing and explain its purpose.

30. Sketch the symbols used on prints to identify the following:
    1. Countersink ________
    2. Diameter ________
    3. Counterbore ________
    4. Depth ________
    5. Position ________

31. What is the symbol used to identify a datum plane?

32. What GD&T symbol would be used to control a 90-degree relationship between two surfaces?

33. Determine the allowance for an RC3 fit at a nominal diameter of 1.00 inch.

Section 3: Job Planning, Benchwork, and Layout

Unit 2 Layout

Name: ______________________ Date: ______________

Score: ______________ Text pages 247–268

Learning Objectives

After completing this unit, you should be able to:

- Explain the purpose of a layout
- Explain the uses of common semi-precision layout tools
- Explain the uses of common precision layout tools
- Perform typical mathematical calculations required to perform layout
- Explain basic layout procedures

Carefully read the unit, and then answer the following questions.

Multiple Choice

1. Layout is used to
   a. determine the number of pieces that can be made from a piece of stock/material.
   b. help determine where to machine on a piece of stock/material.
   c. help determine the orientation of the workpiece on a piece of stock/material.
   d. All of these choices are correct.
   e. None of these choices is correct.

2. What is a tool that can be used to create circles, radii, and arcs with dimensions less than 1″?
   a. Center head
   b. Combination set
   c. Trammel
   d. Divider

3. The main difference between a prick punch and a center punch is the
   a. length.
   b. angle of the point.
   c. material.
   d. diameter.

4. What tool can be used to step off equal spaces across a line?
   a. Divider
   b. Hermaphrodite caliper
   c. Protractor
   d. Scriber

Name: ________________________________ Date: ________________

5. Precision layout is considered when tolerances are closer than
   a. 1/64″.
   b. 0.030″.
   c. 0.5 mm.
   d. All of these choices are correct.
   e. None of these choices is correct.
6. To lay out a line parallel to an edge at a given distance, use a
   a. divider.
   b. hermaphrodite caliper.
   c. trammel.
   d. rule and scriber.
7. A surface gage can be set to a desired height using a
   a. sine bar.
   b. parallel.
   c. height gage.
   d. combination square.
8. During layout, a flat reference surface can be provided by a
   a. surface plate.
   b. combination square.
   c. surface gage.
   d. rule.
9. What shape can be constructed inside a circle with a divider that is set to the radius of that circle?
   a. Square
   b. Pentagon
   c. Hexagon
   d. Octagon
10. What tool can be used to keep parts steady during layout?
   a. Angle plate
   b. Combination square
   c. Surface gage
   d. Height gage

True or False

11. Protractors can be used to lay out angles with an accuracy of .01 degrees. ________
12. Semi-precision layout can be used when print tolerances are 1/32″ or greater. ________
13. A surface gage has the ability to directly set measurements. ________
14. Parallels are used to measure the sizes of workpiece features. ________
15. Details about hazards and handling of layout fluid are found on an NFPA label. ________

Name: ______________________ Date: ______________

Computations

16. What is the largest square that could be produced from 1.5″ diameter round material? (Round answer to 2 decimal places.)

____

____

17. A 1″ square piece of material is needed, but only round material is available. What size piece of round material would be required to produce this 1″ square? (Round answer to 3 decimal places.)

____

____

18. What divider setting is needed to construct a 1.25″ diameter circle?

____

____

Short Answer

19. What is the purpose of layout?

____

____

20. What semi-precision tool can be used to find the center of a cylindrical piece of material?

____

____

21. What layout tool would be a good choice for laying out a 44″ diameter?

____

____

22. What advantages does a height gage (vernier, dial, or digital) have over a surface gage for layout?

____

____

23. What purpose do parallels and setup blocks serve?

____

____

24. What type of tool can be used with gage blocks and a surface plate to position work for very accurate angular layout?

____

____

25. List two uses of a V-block.

____

____

26. What is the difference between a center punch and a prick punch?

____

____

Name: ____________________ Date: ____________

27. Describe a minimum of three general guidelines to follow when performing a layout.

____

____

Unit 3 Hand Tools

Name: ______________________ Date: ______________

Score: ______________ Text pages 269–298

Learning Objectives

After completing this unit, you should be able to:

- Describe safety rules for hand tools
- Describe common screwdriver types
- Describe common types of pliers
- Describe common hammer types
- Describe common wrench types
- Identify clamping methods
- Describe hacksaw use
- Describe file use
- Describe common abrasives
- Describe the purpose of presses

Carefully read the unit, and then answer the following questions.

Multiple Choice

1. Which would be the best hammer to prevent damaging an aluminum workpiece?
- **a.** Brass head
- **b.** Ball peen
- **c.** Soft face
- **d.** Dead blow

2. When seating a workpiece in a vise on parallels, which is the best hammer for the task?
- **a.** Brass head
- **b.** Ball peen
- **c.** Soft face
- **d.** Dead blow

3. Your task is to remove nuts from bolts that have been in place for many years; the heads are corroded and dirty. Which is the best wrench to remove the nuts from the bolts?
- **a.** Hex wrench
- **b.** Box-end wrench
- **c.** Open-end wrench
- **d.** Adjustable wrench

Name: ______________________________ Date: ______________

4. When using an adjustable wrench, the wrench is rotated in the direction of the
   a. adjustable jaw.
   b. solid jaw.
   c. wrench handle.
   d. center line of the bolt head.
5. A __________ fits on a ratchet handle.
   a. Box wrench
   b. Socket
   c. Open end wrench
   d. Spanner wrench
6. A combination wrench has
   a. 2 open ends.
   b. 2 box ends.
   c. 1 open end and 1 box end.
   d. None of these choices is correct.
7. To create a smooth flat surface, use a
   a. single cut file.
   b. double cut file.
   c. curved tooth file.
   d. rasp tooth file.
   e. None of these choices is correct.
8. Wrenches and sockets are generally available in _______ styles.
   a. 4 and 6 point
   b. 6 and 12 point
   c. 6 and 8 point
   d. 8 and 12 point
9. A file should be cleaned with
   a. a file card.
   b. a rag.
   c. compressed air.
   s. a scriber.
10. The coarsest abrasive grit size is
    a. 120.
    b. 180.
    c. 220.
    d. 300.
11. Removing sharp edges from a workpiece is called
    a. deburring.
    b. filing.
    c. polishing.
    d. shaving.

Name: ______________________ Date: ______________

12. A hacksaw cuts on the
  a. back stroke.
  b. front stroke.
  c. back and front stroke.
  d. None of these choices is correct.

True or False

13. The larger the screwdriver, the more force you can apply to the screw. ______
14. Box-end wrenches produce the same force on a bolt as open-end wrenches. ______
15. A vise with serrated hard jaws should be used for clamping small parts. ______
16. A file cuts on both the front stroke and back stroke. ______
17. It is safe to carry screwdrivers in pockets. ______
18. A hook spanner has pins that fit in holes in the face of a nut. ______
19. An arbor press creates much more force than a hydraulic press of the same size. ______

Identification Matching

20. Select the best hand tool from the list for the given application.

a. Form a sheet metal rivet	______	1. Dead blow hammer
b. Remove a six-point spline screw	______	2. Ball peen hammer
c. Clamp a large, round workpiece while keeping the jaws parallel	______	3. Tongue and groove pliers
d. Remove a clip lodged between two surfaces	______	4. Diagonal pliers
e. Seat a workpiece in a vise on parallels	______	5. Needle nose pliers
f. Cut a small wire extending from a surface	______	6. Torx screwdriver

Short Answer

21. What is an advantage of using tongue-and-groove pliers?

____

22. What is the purpose of a dead blow hammer?

____

23. In what situation would an open-end wrench be chosen over a box-end wrench?

____

24. What is the biggest disadvantage of using an open-end wrench?

____

25. Why should there be at least three teeth in contact with the material when hacksawing?

____

26. List the four types of file tooth classifications.

____

Name: ______________________________ Date: ______________

27. List the six levels of file coarseness from roughest to smoothest.

____

28. What is a safe edge and what is it used for?

____

29. Identify and describe various types of screwdrivers.

____

30. Describe the purposes of presses.

____

31. Describe safety precautions when using a hammer.

____

Unit 4 Saws and Cutoff Machines

Name: ______________________ Date: ______________

Score: ______________ Text pages 299–315

Learning Objectives

After completing this unit, you should be able to:

- Identify the various sawing machines used in the machining industry
- Describe band saw safety procedures
- Describe different band saw blade materials
- Define TPI/blade pitch and calculate minimum TPI for material thickness
- Identify the three different tooth patterns and their uses
- Explain blade set and its purposes
- Select proper band saw blade width for sawing given radii
- Identify saw tooth geometry
- Calculate band saw blade length
- Describe the band welding procedure
- Describe blade mounting procedure for the vertical band saw

Carefully read the unit, and then answer the following questions.

Multiple Choice

1. The main difference between a vertical and a horizontal band saw is the
  a. type of blade they use.
  b. type of material they can cut.
  c. orientation of the band in the saw.
  d. degree of accuracy of the cut.

2. What part(s) of a flex back blade is(are) hardened?
  a. Teeth only
  b. Body only
  c. Teeth and body
  d. None of these choices

3. The alternate tooth set pattern has teeth that
  a. switch from side to side every other tooth.
  b. are intermittent, with a neutral space between.
  c. are grouped in a wavy pattern.
  d. None of these choices is correct.

Name: ______________________ Date: ______________

4. The metal chip is formed into curls when sawing at what part on the blade?
   a. Tooth tip
   b. Set gap
   c. Gullet
   d. Crest
5. Speeds for band sawing machines are given in
   a. SFPM.
   b. IPM.
   c. SIPM.
   d. RPM.
6. Which blade type can be run at the highest speeds?
   a. Flex back carbon steel
   b. Hard back carbon steel
   c. Bimetal
   d. Carbide tooth
7. To cut hardened steels, use a
   a. chop saw.
   b. cold saw.
   c. band saw.
   d. power hacksaw.
8. Which blade set is commonly used to cut soft materials such as aluminum?
   a. Alternate
   b. Raker
   c. Wavy
   d. Straight
9. After welding a blade, it becomes brittle and must be
   a. flexed.
   b. cooled.
   c. annealed.
   d. quenched.

True or False

10. Thinner materials require finer pitch saw blades. ________
11. Use a push stick for small parts on a vertical band saw. ________
12. Wider band saw blades can cut smaller radii than narrower blades. ________
13. Positive-rake teeth are generally used for softer materials. ________
14. Never hold material by hand when using a horizontal band saw. ________
15. Vertical band saw guides should be installed before tightening the blade. ________
16. A coarser blade should be run at a higher speed than a finer blade. ________

Name: ______________________________ Date: ______________

Computations

17. What minimum number of teeth per inch should be used to saw 3/16″-thick material?

____

____

18. What length of blade stock would be needed for a saw with 24″-diameter wheels that are 36″ on center? (Round answer to nearest whole number.)

____

____

19. What width of saw blade would be needed to saw a contour with a 1/2″ radius? (Refer to Figure 3.4.20 on the textbook page 277.)

____

____

20. What is the minimum blade TPI required to saw 1/4″ thick material?

____

____

Short Answer

21. What is the major disadvantage of the power hacksaw?

____

____

22. What can be done to make turning corners easier when contour sawing?

____

____

23. What is a variable-pitch saw blade, and what are its benefits?

____

____

24. What is blade gage?

____

____

25. What is a blade set, and what is its purpose?

____

____

26. List the three common tooth sets.

____

____

Name: ______________________ Date: ______________

27. Identify and describe three different tooth patterns and their typical uses.

____

____

28. Describe common blade materials and key factors influencing their selection.

____

____

29. Outline the key steps in the band welding process.

____

____

Unit 5 Offhand Grinding

Name: ____________________ Date: ____________

Score: ____________ Text pages 316–325

Learning Objectives

After completing this unit, you should be able to:

- Explain the purposes for offhand grinding
- Select the correct grinding wheel for the operation to be performed
- Identify different types of offhand grinding machines
- Describe the process of mounting a grinding wheel
- Describe wheel dressing and why it is performed
- Explain how to set up a pedestal grinder for safe operation
- Describe safe offhand grinding procedures

Carefully read the unit, and then answer the following questions.

Multiple Choice

1. Before mounting a new grinding wheel on a grinder, what should be done first?
  a. Set the tool rest.
  b. Check grinder safety guards.
  c. Dress the wheel.
  d. Perform a ring test on the wheel to be mounted.

2. Tool rest and spark breaker should be adjusted to a clearance of
  a. 1/16″ from the wheel.
  b. 1/8″ from the wheel.
  c. 5 mm from the wheel.
  d. 1/4″ from the wheel.

3. The process of removing loading and glazing of a grinding wheel is called
  a. pinning.
  b. dressing.
  c. filing.
  d. truing.

4. The type of abrasive used to grind ferrous metals is
  a. ferrous oxide.
  b. diamond impregnated.
  c. aluminum oxide.
  d. green silicon carbide.

Name: ______________________ Date: ______________

5. A wheel should be dressed when it is
   a. glazed.
   b. loaded.
   c. worn unevenly.
   d. All of these choices are correct.
6. If a grinding wheel is cracked,
   a. have it repaired.
   b. mount it on the grinder.
   c. discard it.
   d. None of these choices is correct.
7. An abrasive belt/disc machine can be used to deburr
   a. ferrous metals.
   b. non-ferrous metals.
   c. ferrous and non-ferrous metals.
   d. carbide.
   e. All of these choices are correct.

True or False

8. A 60-grit grinding wheel is coarser than a 120-grit grinding wheel. ________
9. Abrasive belt machines usually remove the material more quickly than grinders. ________
10. The maximum gap for a tool rest should be 1/16″. ________
11. A star dresser can be used when a grinding wheel is glazed. ________
12. Aluminum oxide wheels can be used for grinding ferrous and non-ferrous materials. ________
13. Abrasive belts are used for lighter deburring tasks. ________
14. If a wheel is rated at 3200 RPM and a grinder runs at 3000 RPM, the wheel is safe to use on the grinder. ________

Short Answer

15. What is the purpose of a diamond impregnated wheel, and which type of offhand grinding machine is it typically used with?

____

____

16. Briefly define a loaded grinding wheel.

____

____

17. Briefly define a glazed grinding wheel.

____

____

Name: ______________________ Date: ______________

18. Briefly list five offhand grinding safety precautions.

____

____

19. List three factors when determining the correct grinding wheel size.

____

____

20. Why do you stand to one side of the pedestal grinder during start-up?

____

____

21. How close should a tool rest and spark breaker be set to the wheel on pedestal grinder?

____

____

22. Briefly describe the process of mounting a grinding wheel on a pedestal grinder.

____

____

23. Briefly identify different types of offhand grinding machines.

____

____

Unit 6 Drilling, Threading, Tapping, and Reaming

Name: ______________________________ Date: ______________

Score: ______________ Text pages 326–348

Learning Objectives

After completing this unit, you should be able to:

- Explain benchwork drilling operations
- Explain the purpose of countersinking/chamfering, spotfacing, and counterboring operations
- Identify various reamer types
- Explain uses of different reamer types
- Describe standardized thread systems and their designations
- Identify various types of taps
- Explain uses of different types of taps
- Define TPI, pitch, major diameter, minor diameter, and pitch diameter, root, and crest
- Select correct tap drills for tapping operations
- Identify various types of thread-cutting dies
- Explain the uses of different types of thread-cutting dies
- Calculate the number of turns of a tap and die to produce a given depth and length of thread
- Describe tap removal techniques

Carefully read the unit, and then answer the following questions.

Multiple Choice

1. A machining process that creates a flat spot on a rough surface is
  a. countersinking.
  b. spotfacing.
  c. counterboring.
  d. chamfering.

2. A machining process that creates an included "V" angle or taper in a hole is
  a. countersinking.
  b. spotfacing.
  c. counterboring.
  d. spot drilling.

Name: ______________________________ Date: ______________

3. Center drills are used to
   a. slightly enlarge an existing hole.
   b. create a hole through a workpiece.
   c. create a starting spot for a hole.
   d. All of these choices are correct.
   e. None of these choices is correct.
4. When drilling for a 1/4″ reamer, the proper amount of material to leave for the reamer is
   a. 0.005″.
   b. 0.010″.
   c. 0.015″.
   d. 0.025″.
5. What is the series of the following thread: 3/4-10 UNC -3A?
   a. 3
   b. A
   c. UN
   d. C
6. Most tap drill charts suggest a hole size that will produce about what percentage of thread depth?
   a. 60%
   b. 50%
   c. 80%
   d. 75%
7. Which tap is considered a general, all-purpose tap?
   a. Bottoming tap
   b. Plug tap
   c. Taper tap
   d. Spiral tap
8. What operation can create a very accurate hole size with a very good surface finish?
   a. Countersinking
   b. Chamfering
   c. Drilling
   d. Reaming
9. After drilling, and before tapping, a hole should be
   a. reamed.
   b. countersunk.
   c. counterbored.
   d. spotfaced.
10. What type of tap pulls chips out of the hole being threaded?
   a. Form tap
   b. Hand tap
   c. Spiral flute tap
   d. Spiral point tap

Name: ______________________________ Date: ______________

11. How many turns of a die would be needed to cut an external 3/8-16 to a length of 7/8″?
   a. 13
   b. 13.5
   c. 14
   d. 14.5
12. A M8 × 1.25 hole is currently tapped to a depth of 6.5 mm. How many more turns of a tap are needed to reach a depth of 12 mm?
   a. 4.4
   b. 5.2
   c. 9.6
   d. 15

True or False

13. A counterbore pilot should be slightly larger than the existing hole. ________
14. Reamer RPM should be faster than drilling. ________
15. A 3/8″ diameter hole should be drilled to tap a 3/8-16 thread. ________
16. To thread with a 3/8-16 die, the material diameter should be 3/8″. ________
17. A form tap pulls chips out of the hole being tapped. ________
18. A die is used to create external threads. ________
19. The pitch of a M3 × .5 thread is 3 mm. ________

Identification Matching

20. Identify the following thread parts by putting the corresponding letter in the appropriate circle on the diagram on the following page.
   a. Major diameter
   b. Minor diameter
   c. Root
   d. Crest
   e. Pitch diameter
   f. Included angle

Name: ____________________ Date: ____________

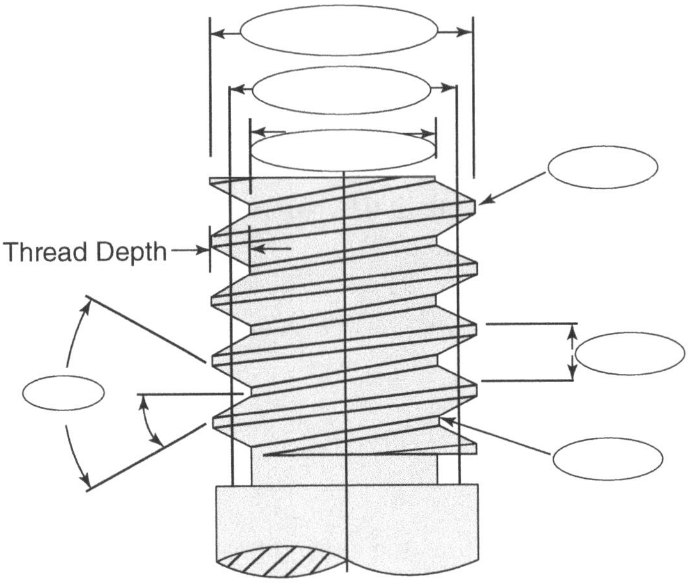

Computations

21. Using the tap drill chart within this unit, determine the tap drill sizes for the cut taps:

a. 3/8-16 UNC ________

b. 1/4-28 UNF ________

c. 10-24 UNC ________

d. 6-40 UNF ________

e. 1/2-13 UNC ________

f. 3/8-24 UNF ________

Short Answer

22. What are the two main standardized systems of thread designations?

____

____

23. What is the pitch of a 1/2-20 thread?

____

____

24. What are the three tap chamfers?

____

____

25. What type of tap pushes chips ahead of the tap?

____

____

Name: ______________________ Date: ______________

26. When drilling, how should feed be adjusted when breaking through a workpiece?

27. What is a blind hole?

28. List the four methods of identifying twist drill diameters.

29. What do UNC and UNF mean?

30. Identify different reamer types and their uses.

31. Describe three broken tap removal techniques.

32. Identify four types of thread-cutting dies.

Section 4: Drill Press

Unit 1 Introduction to the Drill Press

Name: ______________________________ Date: ____________________

Score: ____________________ Text pages 349–356

Learning Objectives

After completing this unit, you should be able to:

- Identify basic types of drill presses
- Identify the major components of the drill press and their functions

Carefully read the unit, and then answer the following questions.

Multiple Choice

1. The spindle speed of a drill press is expressed in what unit?
  a. FPM
  b. RPM
  c. IPM
  d. SFPM

2. A drill press that does not have an automatic feed mechanism is sometimes referred to as a(n)
  a. upright drill press.
  b. bench drill press.
  c. radial-arm drill press.
  d. sensitive drill press.

3. The size of an upright drill press is determined by the
  a. largest size of drill the spindle can hold.
  b. size of the motor.
  c. largest-diameter workpiece that can be drilled on center.
  d. width of the table.

4. What drill press would be a good choice to perform drilling on a large, irregularly shaped casting?
  a. Gang drill press
  b. Radial-arm drill press
  c. Micro drill dress
  d. Sensitive drill press

5. The part of an upright drill press that contains the spindle and is moved up and down by the feed handle is called the
  a. quill.
  b. head.
  c. depth stop.
  d. elevating crank.

Name: ______________________________ Date: ______________

6. Micro precision drill presses are only capable of producing holes
   a. larger than 0.250" in diameter.
   b. smaller than 0.250" in diameter.
   c. that need an accurate location.
   d. in soft material.
7. The movement of a drill press cutting tool into the workpiece is called
   a. speed.
   b. feed.
   c. RPM.
   d. FPM.
8. Speed is changed while the spindle is running on a drill press with
   a. variable speed.
   b. geared head.
   c. step pulley.
   d. All of these choices are correct.
9. Speed is changed while the spindle is running on a drill press with
   a. variable speed.
   b. geared head.
   c. step pulley.
   d. All of these choices are correct.

True or False

10. The size of an upright drill press is given by the largest diameter tool it can hold. ________
11. An upright drill press with a power feed can be called a sensitive drill press. ________
12. The spindle should be running when selecting speeds on a gear-head upright drill press. ________
13. The gang drill press can be set up to perform multiple operations. ________
14. On a radial-arm drill press, the head moves over the secured workpiece. ________
15. Work can be mounted directly to the base of an upright drill press. ________
16. The quill and spindle of an upright drill press are components of the head. ________
17. Spindle speed is measured in FPM, or feet per minute. ________
18. The base and column of an upright drill press are parallel to each other. ________

Identification Matching

19. Match the following upright drill press parts to the function they perform.

a. Column	________	1. Contains motor and belts or gears to set speeds
b. Spindle	________	2. Automatically advances the cutting tool
c. Speed selector	________	3. Used to manually advance the tool into the workpiece
d. Quill feed lever	________	4. Supports the worktable and head
e. Depth stop	________	5. Holds the cutting tool
f. Table clamp	________	6. Adjusts the spindle speed
g. Head	________	7. Secures and locks the table
h. Power feed	________	8. Limits quill travel

Name: ______________________________ Date: ______________

20. Identify the following upright drill parts by putting the corresponding letter in the appropriate circle on the diagram below.
   a. Worktable
   b. Head
   c. Base
   d. Column
   e. Speed selector
   f. Quill feed handle
   g. Spindle
   h. Depth stop
   i. Forward/Off/Reverse switch
   j. Elevating crank
   k. Table clamp
   l. Quill

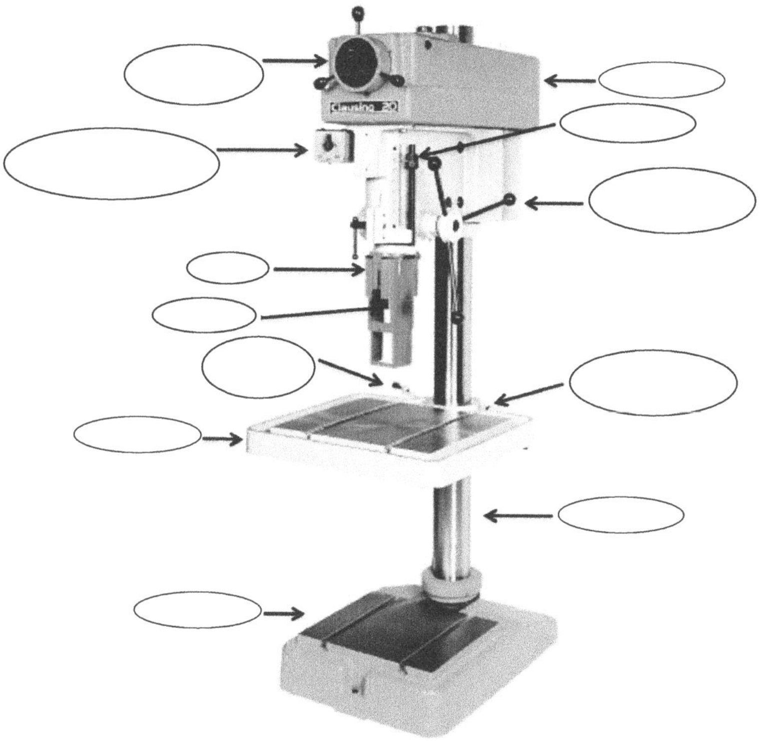

Short Answer

21. What are two methods for advancing a drill press cutting tool into a workpiece?

____

____

Name: ______________________________ Date: ______________

22. List two disadvantages of a step pulley belt-driven upright drill press.

____

____

23. Describe the advantages of using a gang drill press.

____

____

24. Briefly describe the term sensitive drill press.

____

____

25. Briefly describe a micro drill press and its purpose.

____

____

26. What would be the best drill press to machine holes on a very large workpiece?

____

____

27. What would be the best drill press to machine holes as small as 0.005″ in diameter?

____

____

28. List five common machining operations that can be performed on an upright drill press.

____

____

Unit 2 Tools, Toolholding, and Workholding for the Drill Press

Name: ____________________ Date: ____________

Score: ____________ Text pages 357–375

Learning Objectives

After completing this unit, you should be able to:

- Identify the major parts of a twist drill
- Explain the function of the major parts of a twist drill
- Explain various toolholding and workholding devices used on the drill press
- Identify which type of toolholding and workholding devices should be used in various situations

Carefully read the unit, and then answer the following questions.

Multiple Choice

1. The cutting action of a twist drill occurs on what part of the drill?
  a. Dead center
  b. Heel
  c. Flutes
  d. Lips

2. To create a start point for drilling, use a
  a. combination drill and countersink/center drill.
  b. countersink.
  c. twist drill.
  d. reamer.

3. The included angle of a typical twist drill point is
  a. 130 degrees.
  b. 60 degrees.
  c. 82 degrees.
  d. 118 degrees.

4. The part of the twist drill that gives the drill its diameter is called the
  a. body.
  b. flute.
  c. margin.
  d. web angle.

Name: ______________________________ Date: ______________

5. Using the Morse taper numbers given, which of the following drills is the largest?
   a. 4
   b. 5
   c. 6
   d. 7
6. Identify the part of the reamer that actually does the cutting of the material.
   a. Lips
   b. Flute
   c. Chamfer
   d. Body
7. The best type of drill chuck to hold very small drills is the
   a. Morse taper sleeve.
   b. Pin vise chuck.
   c. Jacobs chuck.
   d. keyless chuck.
8. A V-block offers more holding power for cylindrical parts than a vise because of the
   a. clamp.
   b. V-shaped design.
   c. three contact areas.
   d. greater force that can be applied.
9. To remove a Morse taper shank tool from a drill press spindle, use a
   a. wedge.
   b. drift.
   c. key.
   d. tang.
10. What type of reamer has different diameters that can be mounted on one shank?
    a. Shell reamer
    b. Arbor reamer
    c. Sleeve reamer
    d. Shank reamer
11. How can a twist drill be tested after resharpening?
    a. Measuring the diameter across the margins
    b. Drilling a hole in a sample workpiece
    c. Measuring the angle of the point
    d. Measuring the length of the dead center
12. The pilot of a counterbore should be
    a. .003″–.005″ larger than the existing hole.
    b. .003″–.005″ smaller than the existing hole.
    c. .0003″–.0005″ larger than the existing hole.
    d. .0003″–.0005″ smaller than the existing hole.

Name: ______________________________ Date: ______________

13. To drill a hole through a ½″ thick plate measuring 15″ × 15″, secure the plate using
   a. a vise.
   b. an angle plate.
   c. a v-block.
   d. hold-down clamps.
14. Use a sleeve to mount a #3 Morse taper shank drill in a drill press with a
   a. #1 Morse taper spindle.
   b. #2 Morse taper spindle.
   c. #3 Morse taper spindle.
   d. #4 Morse taper spindle.
15. The three main parts of a twist drill are the
   a. point, body, and shank.
   b. point, body, and tang.
   c. tip, body, and shank.
   d. tip, body, and tang.

True or False

16. Parallels can be used to support the workpiece and provide clearance between the table. ________
17. Morse taper cutting tools can be mounted in drill chucks. ________
18. Drills with diameters larger than 1/2″ must have a straight shank. ________
19. Countersinks are available with many different included angles. ________
20. A Morse taper sleeve reduces the size of a tool's shank taper. ________
21. A spotface is machined deeper than a counterbore. ________
22. When removing a taper shank drill from the spindle, the radius edge of the drift should be placed against the tang. ________

Identification Matching

23. Match the following twist drill body parts to their appropriate function.

a. Lips	________	1. Tapered portion that makes up the centermost part of the drill
b. Chisel edge	________	2. Provides area for mounting the drill bit
c. Flutes	________	3. Pathway for chips and coolant
d. Margin	________	4. Angle of the spiral relative to the center axis
e. Body clearance	________	5. Only part of the drill that does cutting
f. Helix angle	________	6. Prevents rubbing of the drill in the hole
g. Web	________	7. Point where the cutting lips meet at the center of the drill
h. Shank	________	8. Thin strip along the edge of the flute

Name: ______________________________ Date: ______________

Short Answer

24. What two materials can cutting tools be made from?

____

____

25. What are the two types of drill shanks?

____

____

26. What are the three purposes of the flutes of a twist drill?

____

____

27. Name the three primary sections of the drill bit.

____

____

28. Briefly explain the use of a Morse taper sleeve and socket.

____

____

29. What are the two main reasons to use a countersink?

____

____

30. What is the purpose of a pilot on a counterbore?

____

____

31. List three of the available included angles of countersinks.

____

____

32. What are the different types of drill lengths, and how are they used in practice?

____

____

Unit 3 Drill Press Operations

Name: ______________________________ Date: ______________

Score: ______________ Text pages 376–394

Learning Objectives

After completing this unit, you should be able to:

- List drill press safety procedures
- Define cutting speed, RPM, and feed
- Perform speed and feed calculations for holemaking operations
- Describe drilling operations
- Describe reaming operations
- Describe countersinking operations
- Describe counterboring/spotfacing operations
- Describe tapping operations
- Calculate countersink feed depth
- Estimate the number of tap turns to achieve a given thread depth

Carefully read the unit, and then answer the following questions.

Multiple Choice

1. The pilot of a counterbore should be
  a. The same size as the existing hole.
  b. .003–.005″ larger than the existing hole.
  c. 003–.005″ smaller than the existing hole.
  d. None of these choices is correct.

2. Cutting speed is the distance that a point on the circumference of a rotating cutting tool travels in 1 minute. It is stated in what unit?
  a. SFPM
  b. SFM
  c. FPM
  d. All of these choices
  e. None of these choices

3. Drill press feed rates are expressed in what unit?
  a. IPR
  b. SFPM
  c. RPM
  d. MPR

Name: ______________________________ Date: ______________

4. What is the RPM used for a piece of material with a CS of 90 and a drill with a diameter of 1/2″?
   a. 600 RPM
   b. 687 RPM
   c. 500 RPM
   d. 725 RPM
5. Speeds for reaming are about
   a. 20–25% of those used for drilling.
   b. 30–40% of those used for drilling.
   c. 50–60% of those used for drilling.
   d. 150% of those used for drilling.
   e. 200% of those used for drilling.
6. Feed rates for reaming are about
   a. 1/4 of those used for drilling.
   b. 1/2 of those used for drilling.
   c. 1.5 times those used for drilling.
   d. 2 times those used for drilling.
7. What part of the drill press can be used to set tool travel for a counterboring operation?
   a. Quill
   b. Feed handle
   c. Quill depth stop
   d. Spindle
8. The RPM for counterboring and spotfacing should be
   a. 20–30% of drilling speed.
   b. 30–40% of drilling speed.
   c. 40–50% of drilling speed.
   d. 50–60% of drilling speed.
9. If a 3/8″-16 bolt had a requirement to be 5/8″ deep in a workpiece, how many turns would be needed by the tap?
   a. 16
   b. 10
   c. 13
   d. 15
10. Tool breakage is usually caused by
    a. feed rates that are too slow.
    b. feed rates that are too high.
    c. spindle speeds that are too low.
    d. spindle speeds that are too high.
11. HSS stands for
    a. high speed steel.
    b. high standard steel.
    c. high strength steel.
    d. None of these choices.

Name: ______________________________ Date: ______________

12. Creating a starting point for a twist drill is called
    a. locating.
    b. pointing.
    c. centering.
    d. spotting.
13. When drilling holes over ½″ in diameter, first create a smaller hole called a
    a. start hole.
    b. guide hole.
    c. pilot hole.
    d. spot hole.
14. Alternate drilling and retracting is called
    a. peck breaking.
    b. chip breaking.
    c. chip drilling.
    d. peck drilling.
15. How much smaller should a hole be drilled to ream .375″?
    a. .005
    b. .010
    c. .015
    d. .020

True or False

16. When machine reaming a hole, if too little material is left for finishing, the reamer may not cut properly. ________
17. A good practice is to always bottom the reamer in a blind hole to ensure a complete diameter. ________
18. The angle of a drill point is included when specifying the depth of a hole. ________
19. Drill press feed rate is measured in FPR. ________
20. When drilling the same material, RPM will be higher for larger diameter drills. ________
21. Tapping can be performed on the drill press with a tapping head. ________
22. Harder materials require higher cutting speeds. ________
23. Carbide tools can generally be run 2–3 times faster than HSS tools. ________

Computations

24. Calculate spindle RPM for the operation of drilling a 13/16″ hole in CRS, which has a 90-SFPM cutting speed. Round answer to the nearest whole number.

____

____

25. Calculate feed depth for a 90-degree countersink used to machine a 3/4″-diameter countersink on an existing 3/8″-diameter hole.

____

____

Name: ______________________________ Date: ______________

26. If a tapped 1/4-20 hole requires a measured thread depth of 0.4375", how many turns of the tap are required?

27. How do countersinking speeds compare to drilling speeds?

Short Answer

28. Briefly explain feed rate, and how different metals affect its selection.

29. List the two formulas used to calculate spindle RPM.

30. Which material is easier to machine: one with a 55% machinability rating, or one with a 75% machinability rating?

31. List two tools that can be used to center a hole location under the center of the drill press spindle.

32. How does Brinell hardness relate to SFPM?

33. List drill press safety precautions.

Section 5: Turning

Unit 1 Introduction to the Lathe

Name: ______________________________ Date: ______________

Score: ______________ Text pages 395–405

Learning Objectives

After completing this unit, you should be able to:

- Explain the principal operation of a lathe
- Identify the three common types of lathe spindle noses
- Explain the functions of the parts of the lathe
- Explain how lathe size is specified

Carefully read the unit, and then answer the following questions.

Multiple Choice

1. The spindle of a lathe uses an internal ________ to hold and align workholding devices.
   a. drawbar
   b. taper
   c. split nut
   d. collet nut
2. To change speeds (RPM) on a belt-driven headstock with a step-cone pulley, the operator must
   a. stop the machine and change the lever positions.
   b. leave the machine running and change the lever position.
   c. stop the machine and manually change the belt location.
   d. leave the machine running in low and manually change the belt position.
3. Feed on a lathe is referred to in which of the following units?
   a. SFPM
   b. FPM
   c. IPM
   d. FPR
4. What is the orientation of the carriage movement to the bed ways?
   a. Parallel
   b. Perpendicular
   c. Angular
   d. Adjacent
5. What part of the lathe allows angular tool movement of the cutting tool?
   a. Carriage
   b. Cross slide
   c. Tailstock
   d. Compound rest

Name: ________________________________ Date: ________________

6. On a manual lathe, the controls for tool movement are located on the
   a. saddle.
   b. apron.
   c. tailstock.
   d. bed.
7. Bed length is one of the factors that determines the size of a lathe. What is the other factor?
   a. Motor size
   b. Spindle capacity
   c. Swing
   d. Tool capacity
8. The part of the carriage that slides on the bed ways is called
   a. apron.
   b. compound rest.
   c. saddle.
   d. cross slide.
9. The correct time to begin each threading pass is shown on the
   a. thread dial.
   b. feed change knob.
   c. lead screw.
   d. gear box.
10. The device used to compensate for machine wear in the cross-slide dovetail is called a
    a. way shim.
    b. tapered key.
    c. gap wedge.
    d. gib.
11. On some lathes, a section of the bed can be removed to allow for larger diameter workpieces. What is this style of lathe called?
    a. Sectional bed
    b. Gap bed
    c. Extenda-bed
    d. Trough bed
12. The precision cylinder in the tailstock that moves in and out by rotating the handwheel is called a
    a. telescoping mount.
    b. spindle.
    c. feed rod.
    d. quill.

True or False

13. The feed rod is a long shaft, either round or hexagonal, that transmits power to the carriage apron during threading. ________
14. The lathe swing dimension determines the length of the part that can be turned. ________

Name: ______________________________ Date: ______________

15. The cross slide motion can be used to cut chamfers and angles on the workpiece. ________
16. On a geared head lathe, the spindle should be running when changing gears. ________
17. The half-nut lever is used to engage the carriage feed when cutting threads. ________

Identification Matching

18. Identify the parts of the lathe by writing their names in the blanks in the following diagram.
   a. Feed rod
   b. Leadscrew
   c. Spindle control lever
   d. Feed rack

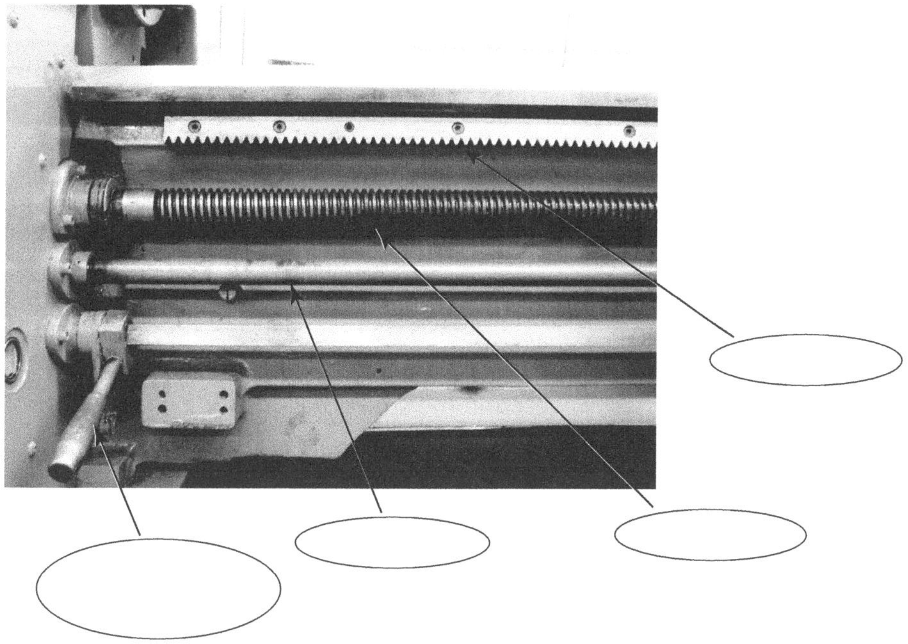

19. Identify the parts of the lathe by writing their names in the blanks on the following diagram.
   a. Feed reverse knob
   b. Spindle clutch lever
   c. Apron hand wheel
   d. Half-nut lever
   e. Thread dial
   f. Feed control clutch
   g. Feed change knob

Name: ______________________ Date: ______________

Short Answer

20. Explain the purpose of a lathe feed rack.

____

____

21. List and briefly describe the three terms used to describe the size of a lathe.

____

____

22. What are two uses of a lathe tailstock?

____

____

23. List five major controls that are on the lathe apron.

____

____

24. Which control would be used to manually feed a cutting tool longitudinally across the workpiece?

____

____

25. Which control would be used to manually move a cutting tool perpendicular to the ways?

____

____

26. List the different types of lathe spindle noses, and identify which type is least commonly used.

____

____

Section 5: Turning

Unit 2 Tools, Toolholding, and Workholding for the Lathe

Name: ______________________________ Date: ______________

Score: ______________ Text pages 406–441

Learning Objectives

After completing this unit, you should be able to:

- Explain the differences between universal-type and independent-type chucks
- Explain the function and application of a three-jaw universal chuck
- Explain the function and application of a four-jaw independent chuck
- Explain the function and application of collets
- Explain the uses of various types of lathe centers
- Explain the use of mandrels
- Explain the applications of a steady rest and follower rest
- Explain the use of various toolholding devices
- Explain basic cutting tools and cutting-tool geometry
- Explain basic designation of carbide inserts and toolholders
- Describe how a tailstock may be used for holding holemaking tools

Carefully read the unit, and then answer the following questions.

Multiple Choice

1. The main difference between a three-jaw chuck and a four-jaw chuck (besides the number of jaws) is the
   - **a.** size of material they can hold.
   - **b.** physical size of the chuck.
   - **c.** jaw tightening screw.
   - **d.** movement of the jaws.
2. Which of the following workholding devices can be used to hold a workpiece off-center?
   - **a.** Collet chuck
   - **b.** Four-jaw chuck
   - **c.** Three-jaw chuck
   - **d.** Drill chuck
3. Which of the following workholding devices is designed so that it can be readily machined to hold a custom-sized workpiece?
   - **a.** Flex collet
   - **b.** Lathe centers
   - **c.** Emergency collet
   - **d.** Faceplate

Name: ____________________ Date: ____________

4. A typical lathe center has an included angle of
   a. 82 degrees.
   b. 90 degrees.
   c. 30 degrees.
   d. 60 degrees.
5. The main advantage of supporting work and turning between centers is
   a. the work can be removed and remounted without losing accuracy.
   b. tool clearance is greatly improved.
   c. eccentric diameters can be easily machined.
   d. None of these choices is correct.
6. The difference between a steady rest and a follower rest is the
   a. third point of contact provided.
   b. position of the rest on the lathe.
   c. mounting location on the machine.
   d. All of these choices are correct.
   e. None of these choices is correct.
7. Taper shank holemaking tools can be held
   a. in a tailstock mounted drill chuck.
   b. in a rocker-type toolholder.
   c. in a 4-jaw chuck.
   d. directly in the tailstock.
8. Work can be mounted through an existing bore using a special precision cylindrical shaft called a
   a. drawbar.
   b. draw tube.
   c. mandrel.
   d. center shaft.
9. What engages with the lathe dog to transfer spindle rotation to the workpiece when holding work between centers?
   a. Drive plate
   b. Faceplate
   c. Three-jaw chuck
   d. Spindle taper
10. A solid tapered mandrel has approximately how much taper per inch of length?
   a. .0001
   b. .0005
   c. .005
   d. .010
11. What is best applied between a tailstock mounted dead center and the workpiece to lessen friction?
   a. Oil
   b. Anti-seize
   c. Lapping compound
   d. Coolant

Name: ______________________________ Date: ______________

12. A taper mounted collet adapter can be removed from the spindle using a
   a. knock-out bar.
   b. cam-release tool.
   c. pry bar.
   d. drawtube.
13. The tool used to drill a tapered hole to accept a lathe center is called a
   a. countersink.
   b. spot drill.
   c. combination center drill and countersink.
   d. taper-seater.

True or False

14. A square workpiece should be held in a 3-jaw chuck. ________
15. A lathe dog is needed when holding a workpiece between centers. ________
16. ANSI has a basic identification system for carbide inserts. ________
17. Large work is commonly secured to faceplates, which can cause workpiece imbalance if run at high speeds. ________
18. A standard universal 3-jaw chuck can repeat to accuracies better than .0005 TIR. ________
19. The jaws on a three-jaw universal chuck move independently. ________
20. A follower rest is bolted to the carriage and travels along the part with the cutting tool. ________
21. A steady rest is mounted to the carriage of the lathe and moves longitudinally with the carriage. ________
22. A collet offers more runout accuracy than a three-jaw chuck. ________

Short Answer

23. What type of chuck can be used to hold an eight-sided workpiece?

____

____

24. List three types of collets that can be machined to the size needed to hold a workpiece.

____

____

25. Name the taper type commonly found in lathe tailstocks.

____

____

26. What are the differences between universal and independent type chucks?

____

____

27. How does a lathe tailstock assist in holding and operating holemaking tools?

____

____

Name: ____________________ Date: ____________

28. List the two types of tailstock centers and briefly describe their differences.

____

____

29. Briefly describe the difference between a quick-change toolholder and an indexable tool post.

____

____

30. Briefly explain basic cutting tools and cutting-tool geometry.

____

____

31. Identify the basic designations of carbide inserts and toolholders.

____

____

Unit 3 Machining Operations on the Lathe

Name: ______________________ Date: ______________

Score: ______________ Text pages 442–476

Learning Objectives

After completing this unit, you should be able to:

- Explain the relationship between depth of cut and diameter reduction
- Compare and contrast roughing and finishing operations
- Explain lathe speed and feed terms, and calculate spindle speeds and machining time
- Describe lathe safety precautions
- Describe the purposes of facing, turning, and shouldering operations
- Explain methods for lathe holemaking operations
- Explain how to use taps and dies to cut threads on the lathe
- Describe the purpose of form cutting
- Explain grooving and cutoff operations
- Explain the purpose and process of knurling

Carefully read the unit, and then answer the following questions.

Multiple Choice

1. A feed rate of .015 IPR means:
   a. The tool would advance .015″ across the work piece during each pass
   b. The tool would advance .015″ in depth during each pass
   c. The tool would advance .015″ per each revolution of the spindle
   d. The tool would advance .015″ per each in-feed pass
2. The three main factors that determine machining time calculations are
   a. material, hardness of material, and tooling.
   b. tooling, machine, and lubricant.
   c. length of part, RPM, and feed.
   d. feed, cutting-tool shape, and hardness of the material.
3. The cutting direction of a left-handed cutting tool is
   a. right to left.
   b. left to right.
   c. either direction.
   d. None of these choices is correct.

Name: ______________________________ Date: ______________

4. Set the height of the cutting tool tip
   a. against the workpiece.
   b. at the center of the workpiece.
   c. slightly above the center of the workpiece.
   d. slightly below the center of the workpiece.
5. The diameter of a workpiece must be machined from 5″ to 2″. The cutting speed is 90 SFPM. How much faster should the RPM be when turning the 2″ diameter than the 5″ diameter?
   a. 85 RPM
   b. 103 RPM
   c. 250 RPM
   d. 90 RPM
6. When filing on a lathe, how should the file be held?
   a. Right hand on handle, left hand on tip for support
   b. Left hand on handle, no hand on tip for support
   c. Right hand on handle, no hand on tip for support
   d. Left hand on handle, right hand on tip for support
7. Cut-off operations should never be performed when holding work
   a. in a three-jaw chuck.
   b. in a collet.
   c. in a four-jaw chuck.
   d. between centers.
8. How much smaller should a hole be drilled to ream a .750″ diameter?
   a. .005
   b. .010
   c. .015
   d. .020
9. Choose the best procedure to produce a square shoulder on the lathe.
   a. Use the carriage to position the tool at the beginning of the shoulder, lock carriage, feed tool across shoulder using the compound rest.
   b. Use the carriage to position the tool at the beginning of the shoulder, lock carriage, feed tool across shoulder using the cross slide.
   c. Set the tool perpendicular to the bed ways, feed carriage to shoulder position, retract tool.
10. Metal chips should be safely cleared from the lathe with
    a. compressed air.
    b. a brush.
    c. a rag.
    d. abrasive cloth.
11. The strongest carbide insert shape is a
    a. circle.
    b. square.
    c. 80-degree rhombus.
    d. trigon.

Name: ______________________ Date: ______________

12. An insert with a shape designation of "V" is a
   a. 35-degree rhombus.
   b. triangle.
   c. 55-degree rhombus.
   d. square.
13. Which is most appropriate for knurling operations?
   a. Low RPM, high feed, no lubrication
   b. Low RPM, high feed, ample lubrication
   c. High RPM, high feed, no lubrication
   d. High RPM, high feed, ample lubrication
14. The relief between the front of the cutting tool and the workpiece, designed to prevent rubbing is called
   a. inscribed circle.
   b. rake angle.
   c. lead angle.
   d. clearance angle.
15. A special cutting tool containing the reverse contour of the desired part shape is called
   a. contour plunge tool.
   b. form cutting tool.
   c. parting tool.
   d. die grooving tool.

True or False

16. Roughing operations use lower cutting speeds than finishing operations. ________
17. Finishing operations use higher feed rates than roughing operations. ________
18. When facing, the tool tip should be set slightly below center. ________
19. Smaller workpiece diameters require slower spindle RPM than larger diameters. ________
20. Lathes and drill presses use the same unit of measure for feed. ________
21. Hold a file handle in the right hand and the file point in the left when filing on the lathe. ________
22. When performing threading on the lathe with taps and dies, the tailstock can be used to align the tap or die to the part. ________

Identification Matching

23. Identify and label the proper turning operations in the following seven pictures.
   a. Boring
   b. Grooving
   c. Knurling
   d. Facing
   e. Reaming
   f. Turning
   g. Drilling

Name: ______________________ Date: ______________

____

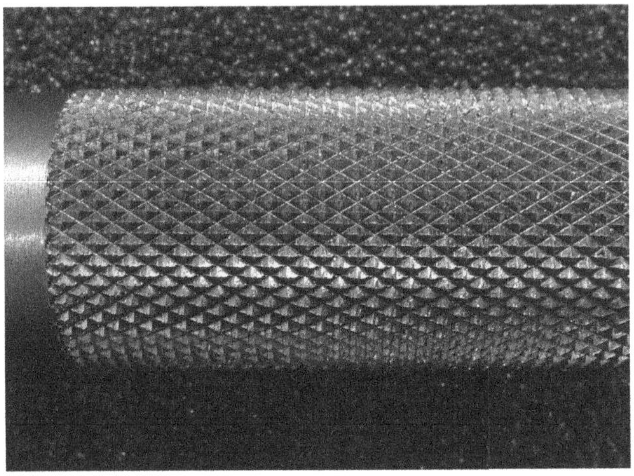

____

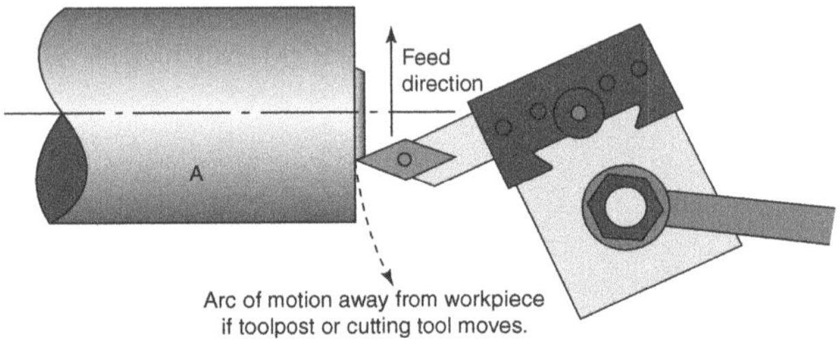
Feed direction
A
Arc of motion away from workpiece if toolpost or cutting tool moves.

____

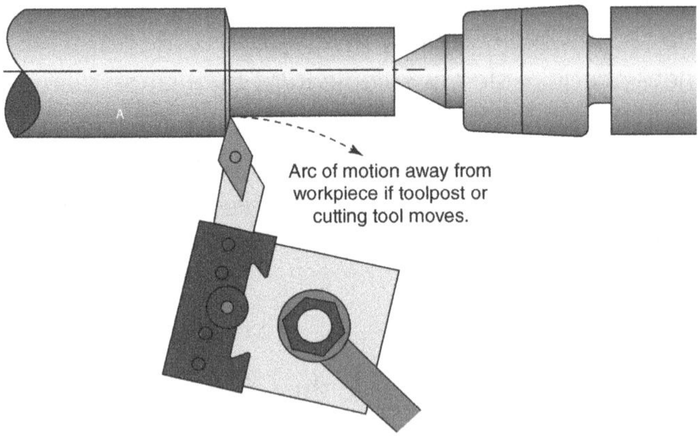
Arc of motion away from workpiece if toolpost or cutting tool moves.

____

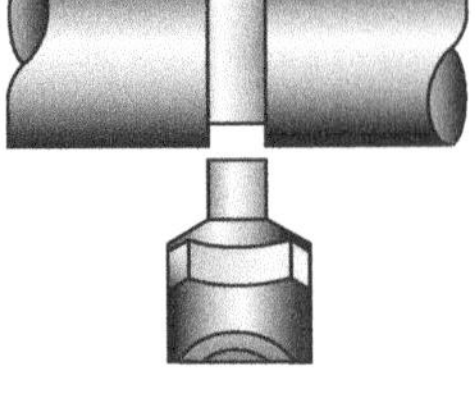

Name: ______________________ Date: ______________

____

____

____

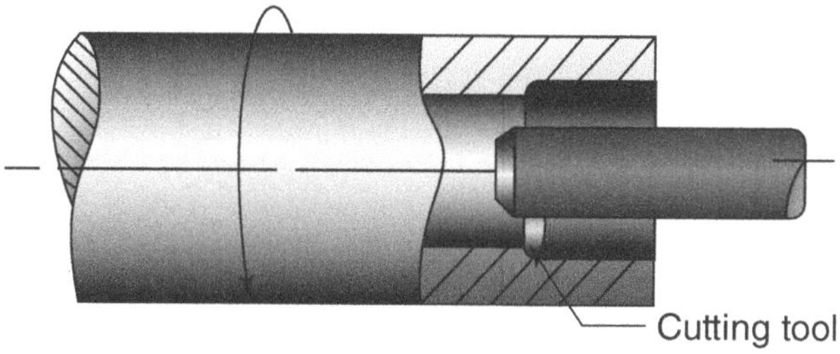

Computations

24. Determine the proper RPM for a 3″ piece of cold-rolled steel with a cutting speed of 90 SFPM. Round answer to a whole number.

____

____

Name: ______________________ Date: ______________

25. Calculate the time to machine an AISI/SAE 4140 steel part 25″ in length and 10″ in diameter using a cutting speed of 110 SFPM and a feed rate of .008 IPR. Round the answer to two decimal places.

Short Answer

26. List 10 safety rules for a lathe.

27. Briefly define boring.

28. List two reasons knurling is performed.

29. Briefly describe grooving and cutoff operations.

30. Briefly explain the relationship between depth of cut and diameter reduction.

Unit 4 Manual Lathe Threading

Name: ______________________________ Date: ______________

Score: ______________ Text pages 477–497

Learning Objectives

After completing this unit, you should be able to:

- Identify the parts of a thread
- Describe the difference between left-hand and right-hand threads
- Describe the difference between single-lead and multiple-lead threads
- Describe the different classes of fit
- Locate thread reference data from charts
- Perform calculations required for thread cutting
- Explain workpiece and tooling setup for thread cutting on the lathe
- Explain the thread cutting process on the lathe
- Explain how to measure thread pitch
- Explain how to measure pitch diameter with a thread micrometer
- Explain how to measure pitch diameter using the three-wire method
- Describe thread forms other than the 60-degree UN form

Carefully read the unit, and then answer the following questions.

Multiple Choice

1. An inch micrometer moves .025″ per one revolution because of this characteristic of the thread.
   - **a.** Pitch diameter
   - **b.** Lead of the thread
   - **c.** Pitch of the thread
   - **d.** Number of threads per inch
2. If a thread had three leads, it would
   - **a.** advance 3 times the rate of a single-lead thread.
   - **b.** advance at the same rate as a single-lead thread.
   - **c.** advance 1/3 the rate of a single-lead thread.
   - **d.** advance at 2 times the rate of a single-lead thread.
3. Which of the characters in the following thread callout signifies whether the thread is internal or external: 11/16″-16 UN 3A?
   - **a.** A
   - **b.** U
   - **c.** N
   - **d.** 3

Name: ______________________________ Date: ____________

4. The main factor of a thread that determines the class of fit is the
   - e. hex angle.
   - a. root size.
   - b. pitch diameter.
   - c. lead of the thread.
5. When cutting a 60-degree thread, the compound rest is set at an angle of
   - a. 29 1/2 degrees.
   - b. 60 degrees.
   - c. 30 degrees.
   - d. None of these choices is correct.
6. The control on the lathe that sets the threads per inch of feed is called the
   - a. headstock.
   - b. feed control level.
   - c. quick-change gear box.
   - d. feed bar.
7. What is the name of the device used to set the threading tool perpendicular to the workpiece?
   - a. Center gage
   - b. Thread dial
   - c. Thread pitch gage
   - d. Thread square
8. What is the name of the device used to determine when to engage the half-nut lever for each threading pass?
   - a. Thread monitor
   - b. Helix counter
   - c. Thread dial
   - d. Tachometer
9. Tapered pipe threads have approximately
   - a. ¼″ of taper per foot.
   - b. 3⁄8″ of taper per foot.
   - c. ½″ of taper per foot.
   - d. ¾″ of taper per foot.
10. Spindle speed for threading should be roughly how much slower than for turning the same diameter?
    - a. One-fourth the speed
    - b. One-half the speed
    - c. Three-quarters the speed
    - d. The same speed
11. Right-hand external threads are created as the
    - a. carriage feeds toward the headstock and the spindle rotates in reverse.
    - b. carriage feeds toward the tailstock and the spindle rotates forward.
    - c. carriage feeds toward the headstock and the spindle rotates forward.
    - d. None of these choices is correct.

Name: ______________________________ Date: ______________

12. Left-hand external threads are created as the
   a. carriage feeds toward the tailstock and the spindle rotates in reverse.
   b. carriage feeds toward the tailstock and the spindle rotates forward.
   c. carriage feeds toward the headstock and the spindle rotates forward.
   d. None of these choices is correct.

13. The loosest fitting UN-thread class is
   a. Class 1.
   b. Class 2.
   c. Class 3.
   d. Class 4.

True or False

14. A triple-lead thread will advance faster than a single-lead thread when rotated. __________

15. A 3B thread requires more care to produce than a 1B thread. __________

16. The compound rest infeed direction determines the correct angle of the thread form. __________

17. The pitch diameter is the key factor for a good-fitting thread. __________

Identification Matching

18. Using the appropriate letter, label the parts of the unified thread shown in the sketch.
   a. Major diameter
   b. Minor diameter
   c. Pitch diameter
   d. Crest
   e. Root
   f. Flank
   g. Thread angle
   h. Helix angle

Name: ______________________________ Date: ______________

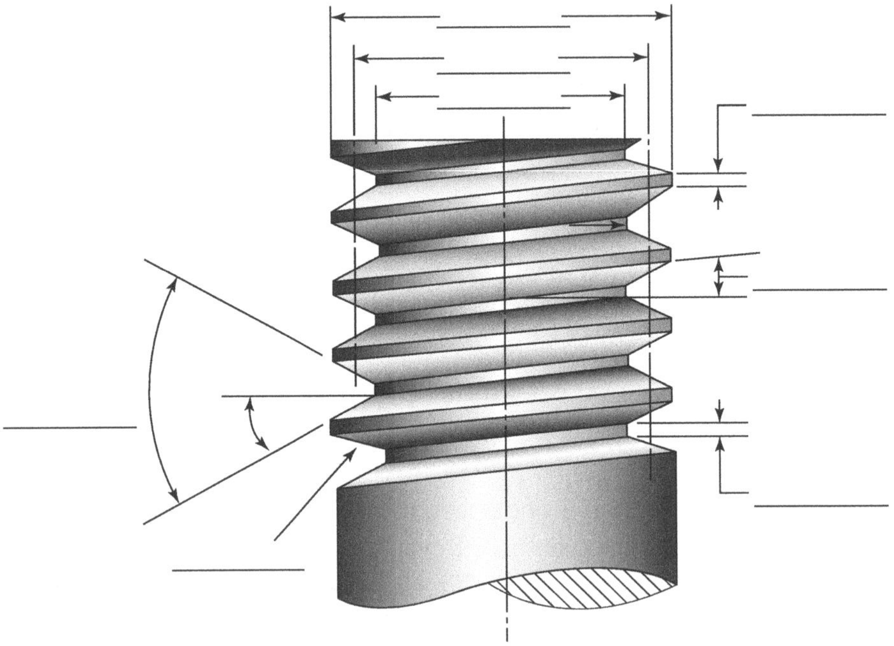

Computations

19. If the quick-change gearbox is set to the setting marked '13' during threading, how many threads per inch will be produced on a 1-inch-diameter shaft?

____

____

20. What are the upper and lower limits for the measurement over wires of a 7/8-14 UNF 3A thread? Use .0500 wires for the calculation. Round answer to four decimal places.

a. upper: ______________________

b. lower: ______________________

21. Refer to the chart in Figure 5.4.6 in the unit to determine the maximum and minimum allowable major diameter of a 3/4-16 UNF 2A. Enter answer with all four decimal places.

____

____

22. What is the estimated compound rest in-feed depth on a 1/4-28 UNF thread? Enter answer with three decimal places.

____

____

Short Answer

23. What type of thread has an asymmetrical form, and what is its purpose?

____

____

Name: ______________________________ Date: ______________

24. What will result when the diameter to be threaded has runout when mounted in the machine?

____

____

25. After making the initial pass when cutting threads, what tool can be used to check for the correct number of threads per inch?

____

____

26. Briefly describe the workpiece and tooling setup for single-point threading on the lathe.

____

____

27. Name three methods that can be used to measure thread pitch diameter when the workpiece is still in the lathe.

____

____

28. Why are acme threads used for power transmission, such as moving the carriage of the lathe?

____

____

29. Briefly describe the difference between left-hand and right-hand threads.

____

____

30. Describe the form of a thread micrometer and its function in measuring the pitch diameter.

____

____

Unit 5 Taper Turning

Name: ______________________________ Date: ______________

Score: ______________ Text pages 498–512

Learning Objectives

After completing this unit, you should be able to:

- Define a taper
- Identify types of dimension specifications for tapers
- Convert taper specifications between TPI, TPF, and angular
- Perform taper calculations for TPI, TPF, and angular specifications
- List methods of turning tapers and their benefits and drawbacks
- Describe setup procedures for taper turning methods

Carefully read the unit, and then answer the following questions.

Multiple Choice

1. Tapers are identified by two different means in the machining industry. One is a "rate of change" specification and the other is
   a. TPI.
   b. TPF.
   c. Angular.
   d. None of these choices is correct.

2. Which of the following methods could produce a 35-degree taper 1.5″ long on a workpiece?
   a. Compound rest
   b. Offset tailstock
   c. Taper attachment
   d. All of these choices
   e. None of these choices

3. Using the chart in Figure 5.5.7, what is the corresponding included angle for 3/8″ TPF?
   a. 2° 59′ 1″
   b. 0° 53′ 43″
   c. 1° 47′ 25″
   d. None of these choices

4. To check a taper attachment setup with the digital read out (DRO) in radius mode, how much cross slide movement should show when the carriage is moved 1″ in the cutting direction?
   a. TPI
   b. One-half of TPI
   c. TPF
   d. One-half of TPF

Name: ______________________________ Date: ______________

5. To reduce machine backlash when using a taper attachment, how should the compound rest be positioned?
   a. Perpendicular to the bed ways
   b. Parallel to the bed ways
   c. 45 degrees rotated clockwise
   d. 45 degrees rotated counterclockwise
6. The spindle speed for taper turning should be calculated based on the
   a. smallest diameter.
   b. largest diameter.
   c. median diameter.
   d. one-half the largest diameter.
7. What type of combination countersink and center drill should be used to drill the part when it will be held using the offset tailstock method?
   a. Plain-type
   b. Bell-type
   c. Involuted
   d. Convoluted
8. A TPI of 3/4″ means that in 2″ of length, the diameter of a taper changes by
   a. 3/8 of an inch.
   b. 3/4 of an inch.
   c. 1.5 inches.
   d. 3 inches.
9. Which taper listed below is a self-holding taper?
   a. Morse
   b. AMT
   c. R-8
   d. 5-C
10. Which taper listed below is a self-releasing taper?
   a. Jacobs
   b. NMTB
   c. Jarno
   d. MT2

True or False

11. A cutting tool can be positioned at a desired angle and plunged into the part to cut a short taper. ________
12. An included angle is measured from one side of the workpiece to the center line. ________
13. Taper attachments can be used for both internal and external operations. ________
14. Compound rests have power feed for cutting tapers. ________
15. The offset tailstock method is suitable for cutting internal tapers. ________
16. A taper attachment is suitable for cutting a 4-degree taper. ________

Name: ______________________ Date: ______________

Identification Matching

17. Identify the best method to cut the given taper on a workpiece.

a. Offset tailstock ________
b. Tool bit formed ________
c. Taper attachment ________
d. Compound rest ________
e. File ________

1. ¼″ TPF, 8″ long taper
2. 30-degree taper, 1.5″ long
3. Chamfer 45 degrees × 1/8″ long
4. .0625 TPI, 16″ long taper
5. Chamfer 1/16″ × 1/16″

Computations

18. If a workpiece needs to be machined with a 2.00″ diameter on one end, a .75″ on the other end, and is 5″ long, what is the TPI? Round answer to two decimal places.

____

____

19. A print specifies a 3-degree included angle. What is the corresponding TPI? Round answer to three decimal places.

____

____

20. Calculate the set over of the tailstock given a 15″ long workpiece and a TPI of 1/32″. Round answer to three decimal places.

____

____

Short Answer

21. List three drawbacks of the offset tailstock method of cutting tapers.

____

____

22. What do TPF and TPI stand for?

____

____

23. What are two drawbacks of using the compound-rest taper-turning method?

____

____

24. How can TPI be converted to TPF?

____

____

25. What are two benefits of the tool bit method of cutting tapers?

____

____

Section 6: Milling

Unit 1 Introduction to the Vertical Milling Machine

Name: ______________________ Date: ______________

Score: ______________ Text pages 513–528

Learning Objectives

After completing this unit, you should be able to:

- Explain the capabilities of a vertical milling machine
- Identify the three axes of vertical milling machine motion
- Identify the components of the vertical milling machine
- Explain the function of the parts of the vertical milling machine
- Describe common optional features of vertical milling machines

Carefully read the unit, and then answer the following questions.

Multiple Choice

1. If the quill is locked, which other part of the mill can be moved to adjust the Z-axis?

a. Head
b. Column
c. Knee
d. Table

2. The taper that is most commonly found in a manual vertical mill spindle is called a(n)

a. #3 Morse Taper.
b. 8-R.
c. R-8.
d. CAT 30.

3. When adjusting the gear range from high to low on a geared head vertical mill, it is important to

a. have the motor running.
b. not have the motor running.
c. have the spindle brake on at all times.
d. have no tools in the spindle.

4. When adjusting the RPM on a variable speed head vertical mill, it is important to

a. have the motor running.
b. not have the motor running.
c. have the spindle brake on at all times.
d. have no tools in the spindle.

Name: ______________________________ Date: ________________

5. What direction does the Z-axis move when standing at a mill in the operating position?
   a. Front/back
   b. Up/down
   c. Left/right
   d. Clockwise/counterclockwise
6. What device is used to secure toolholding devices in the spindle?
   a. Drawtube
   b. Drawbar
   c. Setscrew
   d. Camlock
7. What direction does the X-axis move when standing at a mill in the operating position?
   a. Front/back
   b. Up/down
   c. Left/right
   d. Clockwise/counterclockwise
8. What direction does the Y-axis move when standing at a mill in the operating position?
   a. Front/back
   b. Up/down
   c. Left/right
   d. Clockwise/counterclockwise
9. Micrometer collars on most milling machine cranks are graduated in what increment?
   a. 0.0001″
   b. 0.005″
   c. 0.001″
   d. 0.010″
10. One complete rotation of the table handwheel (or crank) on most machines will result in how much movement?
    a. 0.100″
    b. 0.125″
    c. 0.200″
    d. 0.250″
11. One complete rotation of the knee crank on most machines will result in how much movement?
    a. 0.100″
    b. 0.125″
    c. 0.200″
    d. 0.250″
12. What device can set the limit of quill down-travel so that multiple holes can be drilled quickly and consistently to the same depth?
    a. Quill stop
    b. DRO
    c. Z-Travel restrictor
    d. Post clutch

Name: ______________________ Date: ______________

13. How many degrees can the head tilt side-to-side on a standard knee-mill?
   a. 30
   b. 45
   c. 60
   d. 90

True or False

14. Angles can be cut by tilting the vertical milling machine head. ________
15. The turret movement can be used to machine a radius. ________
16. The power quill feed options for most vertical knee mills are 0.0015″, 0.003″, and 0.006″. ________
17. The table provides left/right movement along the Y-axis. ________

Identification Matching

18. Using the appropriate letter, label the parts of the vertical milling machine in the photo/illustration.
   a. Knee
   b. Saddle
   c. Turret
   d. Head
   e. Table locks
   f. Base
   g. Ram
   h. Table crank
   i. Table
   j. Saddle crank
   k. Knee crank lock
   l. Saddle lock
   m. Column
   n. Knee elevating crank

Name: ______________________ Date: ______________

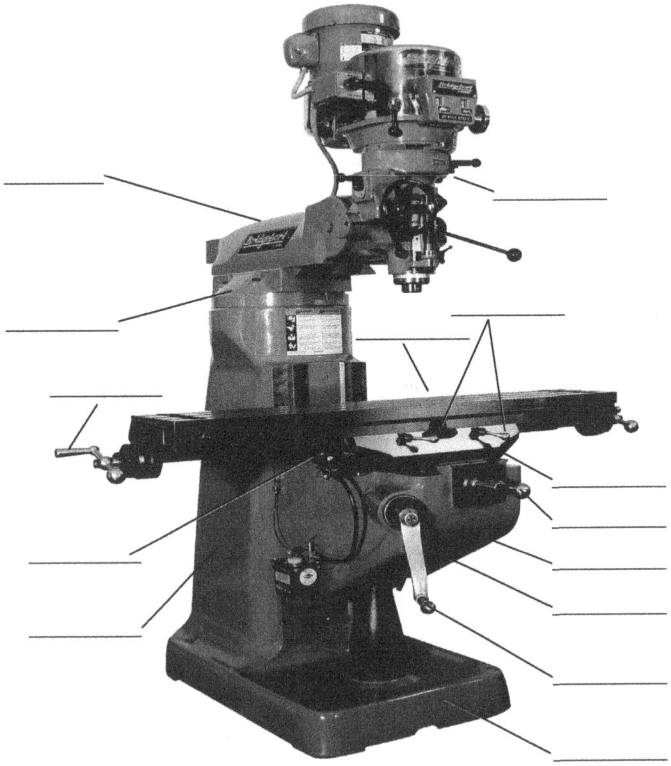

Courtesy of Hardinge Inc.

19. Label the parts of the vertical milling machine head in the photo/illustration.

- **a.** Quill feed selector knob
- **b.** Micrometer adjusting knob
- **c.** High/low range switch
- **d.** Feed control lever
- **e.** Variable speed dial
- **f.** Speed change handwheel
- **g.** Spindle
- **h.** Quill
- **i.** Spindle brake
- **j.** Power feed transmission engagement crank
- **k.** Quill stop
- **l.** High/neutral/low lever
- **m.** Manual feed handwheel
- **n.** Feed reverse knob
- **o.** Quill lock
- **p.** Quill feed handle

Name: ______________________________ Date: ______________

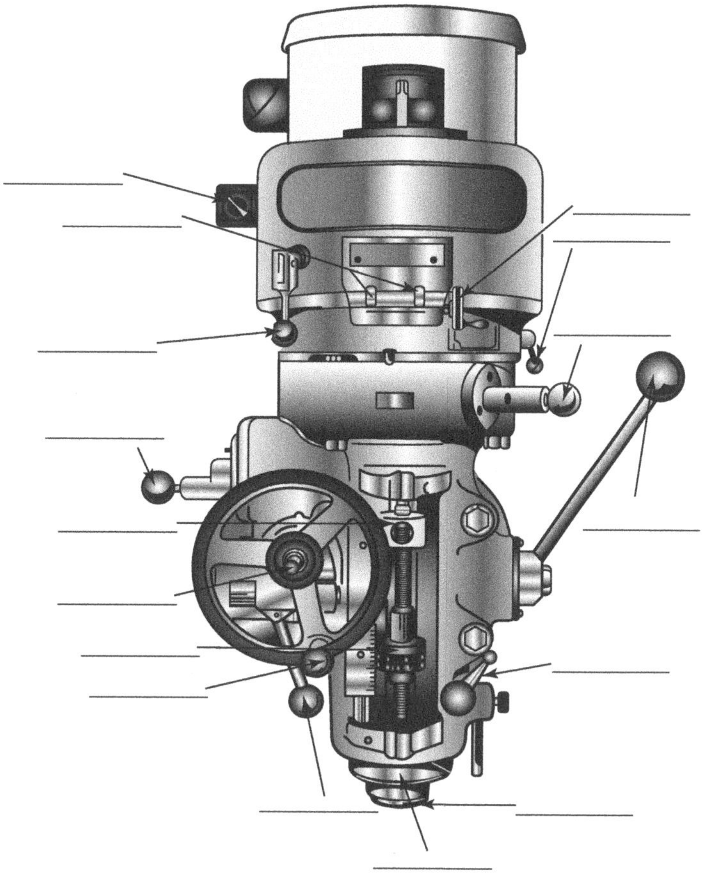

Short Answer

20. Define the term *DRO* and explain how it is used.

____

____

21. Maintenance responsibilities should include supplying lubrication to the sliding surfaces of the vertical milling machine using what type of system? How frequently should this be done?

____

____

22. Briefly explain the purpose of the turret and ram on a vertical knee mill.

____

____

23. What part of the vertical mill is used to control quill feed depth?

____

____

24. Explain the difference between a variable-speed head and a step-pulley milling machine head.

____

____

Unit 2 Tools, Toolholding, and Workholding for the Vertical Milling Machine

Name: ______________________ Date: ______________

Score: ______________ Text pages 529–552

Learning Objectives

After completing this unit, you should be able to:

- Describe the uses of various cutting tools used on the milling machine
- Identify basic designations of carbide inserts
- Explain the uses of various toolholding devices used on the milling machine
- Explain the uses of various workholding devices used on the milling machine

Carefully read the unit, and then answer the following questions.

Multiple Choice

1. Milling cutters and endmills are very brittle and should be stored
   a. to avoid contact with each other.
   b. in plastic or cardboard cases.
   c. without contact with the cutting surface.
   d. All of these choices are correct.
   e. None of these choices is correct.

2. The main difference between a roughing endmill and a finishing endmill is the
   a. length of the cutting edges.
   b. shape of the cutting edges.
   c. size of the endmill.
   d. type of shank.

3. Of the choices below, the best endmill for soft aluminum machining would be a
   a. four-flute.
   b. two-flute.
   c. six-flute.
   d. ball mill.

4. A fly cutter uses how many cutting edges?
   a. One
   b. Two
   c. Three
   d. Four

Name: ______________________ Date: ______________

5. If an internal fillet is needed on a workpiece, what would be the best cutter to use?
   a. Standard endmill
   b. Concave cutter
   c. Radius (bull nose) endmill
   d. Corner rounding endmill
6. The mill vise can be mounted to the table
   a. parallel to the X-axis.
   b. perpendicular to the Y-axis.
   c. offset to the spindle axis.
   d. All of these choices are correct.
   e. None of these choices is correct.
7. A workholding device that is custom-made for a specific workpiece shape is called a
   a. jig.
   b. fixture.
   c. soft table jaw.
   d. custom clamp.
8. The first position of the ANSI carbide insert designation specifies the insert
   a. thickness.
   b. shape.
   c. size.
   d. tolerance.
9. An endmill with a spherical end having a radius that is ½ the size of the cutter's diameter is called a
   a. ballnose endmill.
   b. bullnose endmill.
   c. lollipop cutter.
   d. corner rounding cutter.
10. When using hold-down clamps,
   a. place the stud close to the work and angle the clamp tip downward toward where it contacts the work.
   b. place the stud close to the work and angle the clamp tip downward toward where it contacts the riser block.
   c. place the stud close to the riser block and angle the clamp tip downward toward where it contacts the riser block.
   d. place the stud close to the riser block and angle the clamp tip downward toward where it contacts the work.
11. The fifth position of the ANSI carbide insert designation specifies the insert
   a. thickness.
   b. shape.
   c. size.
   d. tolerance.

Name: ______________________________ Date: ______________

12. The carbide grade best suited for machining stainless steel is
   a. Grade 'P'.
   b. Grade 'H'.
   c. Grade 'S'.
   d. Grade 'M'.
13. Which of the following is a way manufacturers secure carbide inserts to a milling cutter body?
   a. Screw clamp
   b. Wedge clamp
   c. Center-lock screw
   d. Spring action
   e. All of these choices
   f. None of these choices

True or False

14. Vacuum plates are used during heavy milling operations. ________
15. Carbide inserts can be replaced to renew the tool's cutting edges. ________
16. Clearance angle is specified in position three of the ANSI carbide insert designation. ________
17. To mill an internal (fillet) radius, a concave-type endmill can be used. ________
18. Roughing endmills remove material quickly and create smooth surfaces. ________
19. Many drill press holemaking tools can also be used on a vertical mill. ________
20. Slitting saws and form cutters have a straight hole through their centers with a single keyway and are mounted to the spindle with a Morse taper adapter. ________
21. Mill vises may be mounted on a swivel base that allows the vise to be rotated and secured in the desired position. ________

Identification/Matching

22. Circle each endmill below that is capable of center-cutting.

Name: ______________________________ Date: ______________

23. Match the best workpiece holding device to the given workpiece/operation.

a. Special fixture	_____	**1.**	Oddly shaped 8″ × 4″ × 16″ part with largest surface previously machined flat. The other side must be face-milled flat. Only one workpiece must be machined.
b. Magnetic plate	_____	**2.**	1″ square part requiring angled holes. Fifty workpieces must be machined.
c. Vacuum plate	_____	**3.**	Aluminum block 1/2″ × 2″ × 4″ requiring face milling. Ten parts must be machined.
d. Adhesive plate	_____	**4.**	Thin brass sheet metal plates. A light engraving operation must be performed on 200 parts. There are no mounting holes on the parts.
e. Collet fixture	_____	**5.**	10,000 pc production run of large castings having odd shapes and no parallel edges.
f. Fixed milling vise	_____	**6.**	500-piece hole-drilling production run of 10″ × 10″ plastic frames.
g. Angled vise	_____	**7.**	Light milling of 5 iron plates.
h. Toe clamp	_____	**8.**	Thin plastic sheet. Only one workpiece required.
i. Toggle clamp	_____	**9.**	100-part production run of ½″ diameter CRS bar. One flat must be milled on each.

Short Answer

24. List three milling cutters that can be used to machine large flat surfaces.

____

____

25. How would a Morse taper shank drill be held in a typical knee mill?

____

____

26. What are the two names for the type of endmill that has a flat bottom, but a small radius on the corners?

____

____

27. Briefly describe the procedure when securing an endmill in a collet.

____

____

28. List two types of arbor-mounted milling cutters.

____

____

Unit 3 Vertical Milling Machine Operations

Name: ______________________________ Date: ______________

Score: ______________ Text pages 553–597

Learning Objectives

After completing this unit, you should be able to:

- Describe vertical milling machine safety practices
- Explain the process of tramming a vertical milling machine head
- Explain methods of aligning workholding devices and workpieces
- Calculate speeds and feeds for milling operations
- Explain the process of using an edge finder to establish a reference location
- Explain the process of using an indicator to locate the center of a part feature
- Describe the process of boring on a milling machine
- Explain conventional and climb milling
- Explain the process of squaring a block on a milling machine
- List methods for milling angular surfaces
- Explain basic concepts of milling slots
- List methods for milling radii
- Explain the basic steps of milling rectangular pockets

Carefully read the unit, and then answer the following questions.

Multiple Choice

1. The process that is used to adjust the spindle perpendicular to the top surface of the table is called
 a. trimming.
 b. positioning.
 c. tramming.
 d. squaring.

2. The three factors used in the formula for calculating IPM are
 a. RPM, tool diameter, and number of teeth.
 b. RPM, tool length, and number of teeth.
 c. FPT, tool diameter, and number of teeth.
 d. RPM, FPT, and number of teeth.

3. Which of the following methods is used to mill angular surfaces?
 a. Using a swivel-base milling vise
 b. Using concave-rounding milling cutters
 c. Using a bullnose endmill
 d. None of these choices

Name: ______________________________ Date: ______________

4. Choose the appropriate speed for an edge finder from the choices below:
   a. 100 RPM
   b. 1,100 RPM
   c. 500 RPM
   d. 200 RPM
5. If an edge finder with a diameter of 0.200″ is used to locate an edge, how much adjustment is required to bring it to the center?
   a. 0.200″
   b. 0.400″
   c. 0.050″
   d. 0.100″
6. The main difference between conventional milling and climb milling is the
   a. direction of feed.
   b. size of the cutter.
   c. depth of cut.
   d. rate of feed.
7. Choose the appropriate amount of TIR to achieve when aligning a standard milling vise.
   a. 0.00005
   b. 0.0005
   c. 0.002
   d. 0.005
8. When machining a block square, after the first side is machined flat, what should be used to position the part in the milling vise while machining the second surface?
   a. One parallel
   b. Two parallels
   c. A rod and one parallel
   d. A rod and two parallels
9. When multiple parts of the same type are being machined in a vise, what device can be used to efficiently position each one at the same place with an accuracy of 0.001″ to 0.002″?
   a. Scribe mark
   b. Dial indicator
   c. Depth micrometer
   d. Work stop
10. RPM for boring should be
    a. 25% of drilling the same diameter.
    b. 75% of drilling the same diameter.
    c. 100% of drilling the same diameter.
    d. 125% of drilling the same diameter.

Name: ______________________________ Date: ______________

True or False

11. An angled surface can be accomplished by angling the head of the machine. ________
12. An open slot does not pass through the ends of the workpiece. ________
13. When squaring a workpiece, reposition the block for each machining step by placing a newly machined surface against the solid jaw. ________
14. Feed per tooth is also called "chip load." ________
15. When squaring a block, the side with the smallest surface area should be machined first. ________
16. When roughing, use slightly lower speeds, higher feed rates, and deeper cuts. ________
17. Climb milling pulls the workpiece into the cutter. ________
18. Concave milling cutters are used to mill external radii. ________

Computations

19. Calculate RPM and feed rate in IPM for a 3/4"-diameter two-flute HSS endmill cutting aluminum at 250 SFPM with a 0.004" FPT. Round RPM to a whole number and round the feed rate to one decimal place.

____

____

20. Calculate the RPM for boring a 2" diameter hole at 90 SFPM. Round the answer to a whole number.

____

____

Short Answer

21. Define and briefly describe the process of aligning the spindle so it is perpendicular to the milling machine table.

____

____

22. What is the formula for calculating IPM for milling operations?

____

____

23. How much material should be left for a finish pass after rough side (peripheral) milling a surface?

____

____

24. Briefly describe the steps to use an edge finder to locate the edge of a workpiece.

____

____

25. What is the name of the device used for boring on the vertical mill that allows a boring bar to be offset to adjust the size of the hole being machined?

____

____

Name: ______________________________ Date: ______________

26. How would a dial test indicator be used to find the center of an existing hole?

____

____

27. Briefly describe two methods for positioning an endmill in the center of a shaft.

____

____

28. What is a coordinate map?

____

____

29. List three safety precautions that should be observed during any milling operation.

____

____

Unit 4 Indexing and Rotary Table Operations

Name: ______________________ Date: ______________

Score: ______________ Text pages 598–610

Learning Objectives

After completing this unit, you should be able to:

- Describe the capabilities of the rotary table, indexing fixture, and dividing head
- Identify the basic parts of a rotary table, indexing fixture, and dividing head
- Describe basic setup and operation of the rotary table, collet blocks, indexing fixture, and dividing head
- Perform direct and simple indexing calculations

Carefully read the unit, and then answer the following questions.

Multiple Choice

1. Rotary tables are aligned to the center of the spindle with a
  a. dial test indicator.
  b. dial plunge indicator.
  c. dial depth gage.
  d. All of these choices are correct.
  e. None of these choices is correct.

2. Which of the following statements about collet blocks is correct?
  a. They are manually indexed.
  b. The block can be held in a standard milling vise.
  c. They are used to mill a square on a cylindrical workpiece.
  d. All of these choices are correct.

3. The function of a rotary table is to
  a. mill straight lines on large castings.
  b. move the cutter to make arcs and circles.
  c. move the workpiece to make arcs and circles.
  d. mill linear lines on small-diameter workpieces.

4. Using notches on an indexing head to obtain the proper divisions needed is called
  a. simple indexing.
  b. direct indexing.
  c. complex indexing.
  d. spindle indexing.

Name: ____________________ Date: ____________

5. How many turns of a dividing head crank result in 1 turn of the spindle?
   a. 1
   b. 10
   c. 24
   d. 40
6. If a 0.500"-diameter endmill is used to machine the outside of a 4.25"-diameter disk on a rotary table, how much total offset from the center is required?
   a. 2.375
   b. 2.625
   c. 4.5
   d. 4.75
7. If a 0.500"-diameter endmill is used to machine the inside of a 4.25"-diameter hole on a rotary table, how much total offset from the center is required?
   a. 1.625
   b. 1.875
   c. 3.75
   d. 4.0
8. A gear blank requiring 41 gear teeth to be milled is best machined with a
   a. rotary table.
   b. direct indexing fixture.
   c. direct indexing head.
   d. dividing head.
9. A part requiring a ball-milled groove on a 4" radius around 270 degrees of the part is best held with a
   a. rotary table.
   b. direct indexing fixture.
   c. direct indexing head.
   d. dividing head.

True or False

10. The rotary table centerline should be aligned with the spindle before aligning the workpiece on the rotary table with an indicator. ________
11. Parts may be mounted to the rotary table using T-slot hardware. ________
12. If no fractional turns are required, any indexing plate can be used. ________
13. The number of divisions is limited by the number of notches or holes in the direct index plate. ________
14. Collets are used in all indexing heads to hold the workpiece. ________
15. When setting sector arm spacing, the current hole is counted. ________

Name: ______________________________ Date: ______________

Computations

16. Using a direct indexing head with a 24-notch-plate, how many notches must be moved to create 4 divisions?
17. Calculate the number of whole and partial turns using the index plates below for the simple indexing situations:

    Plate 1 has hole circles with 15, 16, 17, 18, 19, and 20 holes
    Plate 2 has hole circles with 21, 23, 27, 29, 31, and 33 holes
    Plate 3 has hole circles of 37, 39, 41, 43, 47, and 49 holes

    a. Make 32 divisions

    ____________ whole turns and ____________ holes in a ____________ -hole circle

    b. Make 39 divisions

    ____________ whole turns and ____________ holes in a ____________ -hole circle

    c. Make 56 divisions

    ____________ whole turns and ____________ holes in a ____________ -hole circle

Short Answer

18. List three uses of a rotary table.
19. What is the main function of a dividing head?
20. What is the difference between the abilities of a rotary table and a dividing head?
21. Briefly describe two methods to align a dividing head with the machine.
22. What is the difference between simple and direct indexing?

Unit 1 Introduction to Precision Grinding Machines

Name: ______________________ Date: ______________

Score: ______________ Text pages 611–618

Learning Objectives

After completing this unit, you should be able to:

- Explain the benefits of precision grinding
- Describe various types of grinders and their capabilities
- Identify basic parts of a surface grinder

Carefully read the unit, and then answer the following questions.

Multiple Choice

1. The primary purpose of a surface grinder is to produce
 a. flat surfaces.
 b. curved surfaces.
 c. angled surfaces.
 d. All of these choices are correct.
 e. None of these choices is correct.

2. Precision grinding can easily produce a surface finish as fine as
 a. 200 grit.
 b. 16 microinches.
 c. 50 grit.
 d. 1 microinch.

3. A typical horizontal spindle surface grinder uses the ______________ of the grinding wheel to remove material.
 a. edge
 b. side
 c. periphery
 d. ID

4. A grinder that performs machining tasks similar to those of a lathe is called a
 a. horizontal grinder.
 b. vertical grinder.
 c. jig grinder.
 d. cylindrical grinder.

Name: ______________________________ Date: ______________

5. Which of the following is NOT a characteristic of a jig grinder?
   a. Has precise movement of the X- and Y-axes
   b. Uses a spindle-mounted grinding wheel
   c. Used for sharpening cutting tools
   d. Produces a smooth finish, accurate hole
6. The elevating handwheel of a surface grinder is typically graduated in what increments?
   a. 0.00001″
   b. 0.0001″
   c. 0.001″
   d. 0.010″
7. Another name for a vertical spindle rotary surface grinder is
   a. cylindrical grinder.
   b. blanchard grinder.
   c. bullard grinder.
   d. woodford grinder.
8. Proper grinding wheel action removes material by
   a. burnishing.
   b. sparking.
   c. rubbing.
   d. cutting.
9. The rotational speed of the workpiece in a centerless grinder is controlled by the
   a. grinding wheel.
   b. work rest.
   c. work brake.
   d. regulating wheel.

True or False

10. The main difference between off-hand grinding and precision grinding is the amount of material removed. ________
11. After setting the amount of travel on the automatic reciprocating surface grinder, engaging the longitudinal feed will continuously cycle back and forth under the wheel. ________
12. Vertical spindle grinders are available in either reciprocating or rotary table styles. ________
13. A tool and cutter grinder is considered an off-hand grinder. ________
14. Surface grinders can be used to remove both ferrous and non-ferrous materials. ________

Name: ______________________________ Date: ______________

Identification Matching

15. Match the given precision grinding machine to the grinding task.

a. Horizontal spindle surface grinders	______	**1.**	Uses a solid cylindrical or cup-shaped wheel to grind a workpiece flat.
b. Vertical spindle surface grinder	______	**2.**	Work is rotated against the rotation of the grinding wheel. Outside or inside diameters can be ground.
c. Cylindrical grinders	______	**3.**	Uses a grinding wheel mounted in a vertical spindle to grind precision holes
d. Centerless grinders	______	**4.**	Used to resharpen cutting tools or to create customized tools from HSS or carbide tool blanks
e. Tool grinders	______	**5.**	Uses the periphery of a solid wheel to grind a workpiece flat
f. Jig grinders	______	**6.**	Grinds outside diameters. The work is supported with a work rest while the part is fed between the grinding wheel and a regulating wheel

Short Answer

16. What are the two reciprocating surface grinder table movements that are similar to movements of the vertical milling machine?

____

____

17. Briefly explain the longitudinal and cross-feed movements while grinding on an automatic reciprocating surface grinder.

____

____

18. List three ways to mount a workpiece for cylindrical grinding.

____

____

19. What is the main function of a jig grinder?

____

____

Unit 2 Grinding Wheels for Precision Grinding

Name: ______________________ Date: ______________

Score: ______________ Text pages 619–625

Learning Objectives

After completing this unit, you should be able to:

- Identify grinding wheel shapes
- Explain the grinding wheel identification system
- List types of abrasives used to make grinding wheels
- Explain grit size (grain size)
- Explain the hardness scale of grinding wheels
- List the different types of grinding-wheel bonding agents
- Explain wheel structure
- Explain grinding wheel characteristics suitable for various applications
- Describe the use of superabrasives for precision grinding

Carefully read the unit, and then answer the following questions.

Multiple Choice

1. Which wheel type is used on tool and cutter grinding machines?
   - **a.** Cup wheel
   - **b.** Saucer wheel
   - **c.** Dish-shaped wheel
   - **d.** All of these choices
   - **e.** None of these choices
2. Which character defines the type of wheel abrasive? 38 A 60 J 8 V BE
   - **a.** A
   - **b.** J
   - **c.** V
   - **d.** BE
3. Which character defines the wheel structure? 38 A 60 J 8 V BE
   - **a.** 38
   - **b.** A
   - **c.** J
   - **d.** 8

Name: ______________________________ Date: ______________

4. Silicon carbide grinding wheels are designated by the letter
   a. A.
   b. B.
   c. C.
   d. S.
5. What is the grit of a 38 A 60 J 8 V BE wheel?
   a. 38
   b. 60
   c. 8
   d. BE
6. The spacing between the individual grains of a grinding wheel is called
   a. bond.
   b. grain size.
   c. structure.
   d. grade.
7. What type of grinding wheels can be used to grind extremely hard materials?
   a. Diamond abrasive
   b. CBN
   c. Superabrasives
   d. All of these choices
   e. None of these choices
8. Which is the most common bonding agent used in precision grinding wheels?
   a. Shellac
   b. Resinoid
   c. Vitrified
   d. Silicate
9. What does CBN stand for?
   a. Cubic Based Node
   b. Cubic Brazed Nub
   c. Cubic Boron Nitride
   d. Cubic Benzilate Nitride
10. The ability of the individual abrasive grains to fracture during grinding to create new, sharp cutting edges is called
   a. friability.
   b. fractation.
   c. granulation.
   d. impact dressing.

Name: ______________________________ Date: ______________

True or False

11. "Aluminum oxide" defines the structure of the grinding wheel. ________
12. The letter E identifies shellac wheels. ________
13. Aluminum oxide and silicon carbide are abrasive types used to make most grinding wheels. ________
14. A 120-grit wheel is finer than a 100-grit wheel. ________
15. Generally, hard wheels are used for grinding hard materials and soft wheels are used for grinding soft materials. ________
16. Grain sizes of grinding wheels range from 8 to 600. ________
17. Wheel grades are identified by the letters through A (hardest) through Z (softest). ________
18. The abrasive designation for CBN is the letter "B". ________

Identification Matching

19. Match the correct bond type to the given definition.

a. Vitrified	________	1. Flexible natural or organic bond; it can deteriorate over time and become brittle and weak
b. Silicate	________	2. Used to make very thin wheels
c. Rubber	________	3. Used for rough off-hand grinding under harsh conditions
d. Resinoid	________	4. Synthetic glass bond
e. Shellac	________	5. Used to grind very thin parts and edges of hard materials

Short Answer

20. Explain the scale for the structure of grinding wheels.

____

____

21. Briefly explain the bond or bonding agent of a grinding wheel.

____

____

22. What is the disadvantage of using superabrasives?

____

____

23. What two types of silicon carbide grades are available, and what material do they grind?

____

____

Unit 3 Surface Grinding Operations

Name: ______________________________ Date: ______________

Score: ______________ Text pages 626–644

Learning Objectives

After completing this unit, you should be able to:

- Describe surface grinder safety procedures
- Explain the basic process of mounting and dressing surface grinder wheels
- Describe the applications of workholding devices for grinding
- Describe the process of grinding parallel, square, and angular surfaces
- Describe the methods for side grinding of vertical surfaces
- Identify common grinding problems and their solutions

Carefully read the unit, and then answer the following questions.

Multiple Choice

1. The cushion placed between the grinding wheel and spindle flange is called a
  a. spacer.
  b. blotter.
  c. shim.
  d. None of these choices is correct.

2. Angular surfaces can be ground by holding the workpiece
  a. on a magnetic sine chuck.
  b. in a magnetic V-block.
  c. on an angle plate with clamps.
  d. All of these choices are correct.

3. The technique used to secure a workpiece that is taller than its length or width is called
  a. cramping.
  b. dogging.
  c. blocking.
  d. chucking.

4. When setting up a single-point diamond dresser, where should it be located in relation to the grinding wheel?
  a. Slightly right of center
  b. Slightly left of center
  c. On center
  d. All of these choices

Name: ______________________________ Date: ______________

5. Which of the following is NOT considered a cause of surface scratches?
   a. No coolant
   b. Dirty coolant
   c. Too much coolant
   d. Loaded grinding wheel
6. An appropriate depth of cut for a rough grinding pass on hardened steel is
   a. 0.0001″–0.0002″
   b. 0.001″–0.003″
   c. 0.004″–0.008″
   d. 0.010″–0.020″
7. Material that is attracted to a magnet is called
   a. ferromagnetic.
   b. entrancable.
   c. polarized.
   d. magnatronic.
8. When rough dressing the wheel with a diamond dresser, how much should the wheel be lowered for each pass?
   a. 0.0002″–0.0005″
   b. 0.001″–0.003″
   c. 0.005″–0.008″
   d. 0.010″–0.015″
9. The removal of the cobalt binder in carbide from grinding with water-soluble coolants is known as
   a. erosion.
   b. wash-out.
   c. leaching.
   d. undermining.
10. Burn marks on the workpiece may be caused by
   a. a structure that is too dense.
   b. too much cut depth.
   c. a wheel that is too fine.
   d. All of these choices are correct.

True or False

11. Dressing a grinding wheel sharpens the abrasive grains. ________
12. A wavy finish could be caused by inadequate dressing/truing. ________
13. If the magnetic chuck is worn, it can be ground to create a new flat workpiece mounting surface. ________
14. A magnetic V-block is often used to hold work at a 30- or 60-degree angle to the chuck surface. ________
15. Always shut the wheel off and let it come to a complete stop before making workpiece adjustments or taking measurements. ________

Name: ______________________________ Date: ______________

16. The grinding wheel should travel completely off the work between cross feed step overs. ______
17. Aluminum oxide and silicon carbide wheels are dressed and trued using a brake-truing device. ______
18. A brake-truing device contains an aluminum oxide wheel mounted on a spindle with an automatic braking system. ______
19. The threads on horizontal surface grinder spindles have right-hand threads. ______

Short Answer

20. How can a non-ferromagnetic material be held directly on a magnetic chuck?

____

____

21. How should the cut depth or step over be adjusted if workpiece burning occurs during surface grinding?

____

____

22. What can cause waviness on a ground surface?

____

____

23. Briefly explain "sparking out" and why it is performed.

____

____

24. What operation should be performed on a grinding wheel before mounting, and why?

____

____

25. Briefly describe the methods for side grinding of vertical surfaces.

____

____

Unit 1 CNC Basics

Name: ______________________________ Date: ______________

Score: ______________ Text pages 645–660

Learning Objectives

After completing this unit, you should be able to:

- Explain the function of basic CNC motion-control hardware
- Explain the Cartesian coordinate system for CNC applications
- Explain the polar coordinate system for CNC applications
- Explain the absolute and incremental positioning systems
- Describe the purpose of G- and M-codes
- Describe CNC word address programming
- Describe the difference between modal and non-modal codes
- Describe a "block" of a CNC program
- Describe types of machine axis motion

Carefully read the unit, and then answer the following questions.

Multiple Choice

1. The main difference between a CNC milling machine and a CNC milling center is the
  a. power knee.
  b. power quill.
  c. automatic tool changer.
  d. MCU.
  e. programmable Z-axis.

2. The best type of axis drive screw for use on a CNC machine is the
  a. acme screw.
  b. buttress screw.
  c. square screw.
  d. ball screw.
  e. round screw.

3. Servo motors are necessary to drive the axes of a CNC because of their ability to
  a. support and move heavy objects.
  b. track how far the axis has moved.
  c. use less power.
  d. run for long periods of time without maintenance.
  e. None of these choices is correct.

Name: ______________________ Date: ______________

4. Movements in the "north" direction are given positive values and movements in the "south" direction are given negative values. What axis is being referenced?
   a. X
   b. Y
   c. Z
   d. A
   e. C
5. In absolute positioning, all locations reference the
   a. absolute point.
   b. reference point.
   c. beginning point.
   d. point of origin.
   e. both c and d.
   f. None of these choices is correct.
6. Which of the following statements best describes the difference between modal and non-modal codes?
   a. Modal codes remain active until canceled or replaced, while non-modal codes apply only to the block in which they appear.
   b. Modal codes are the G-codes, whereas non-modal codes are the M-codes.
   c. Modal codes control linear interpolation, whereas non-modal codes control rapid traverse.
   d. Modal codes are used in circular interpolation, whereas non-modal codes are used in linear interpolation.
7. In CNC programming, the End-Of-Block (EOB) character is what character on a standard keyboard?
   a. Asterisk
   b. Period
   c. Exclamation mark
   d. Semi-colon
8. Choose the feedrate that is most likely to be used by a modern machine during a rapid traverse movement.
   a. 5 IPM
   b. 125 IPM
   c. 1500 IPM
   d. 7000 IPM
9. The coordinate plane of the Cartesian coordinate system is broken up into four pieces called
   a. quadrants.
   b. quarter-planes.
   c. corner zones.
   d. quads.

Name: ______________________________ Date: ______________

True or False

10. A "block" is a sequence of program words written on a single line of a CNC program. ________

11. The rectangular coordinate system is the same as a Cartesian coordinate system. ________

12. A polar coordinate system uses linear distances in both the X- and Y-axes to define locations. ________

13. Incremental positioning references each location from the part origin. ________

14. Modal codes are active until the machine is told to do otherwise. ________

15. A conversational machine control is programmed primarily with a user-written G- and M-code program. ________

16. The servo motor rotates the linear guides to move the machine axes. ________

17. The polar coordinate system identifies positions with an angle and a distance from the origin to a location. ________

Identification Matching

18. Match the codes below to the correct code name

a. G00	________	**1.** Inch unit programming
b. G01	________	**2.** Absolute positioning
c. G02	________	**3.** Circular interpolation (CCW)
d. G03	________	**4.** Coolant on
e. G20	________	**5.** Rapid traverse
f. G90	________	**6.** Program end/return to start
g. M06	________	**7.** Tool change
h. M08	________	**8.** Circular interpolation (CW)
i. M09	________	**9.** Linear interpolation
j. M30	________	**10.** Coolant off

Short Answer

19. Briefly explain the purpose of an encoder.

____

____

20. List four reasons why Acme lead screws are not used in CNC machines.

____

____

Name: ______________________________ Date: ______________

21. What are the two positioning systems used in programming to define a workpiece position?

____

____

22. Explain a G-code.

____

____

23. Explain an M-code.

____

____

Unit 2 Introduction to CNC Turning

Name: ______________________________ Date: ______________

Score: ______________ Text pages 661–675

Learning Objectives

After completing this unit, you should be able to:

- Identify CNC turning machine types
- Identify parts of CNC turning machines
- Identify the machine axes used for turning
- Identify toolholding and tool-mounting devices for CNC turning
- Describe the application of toolholding devices for CNC turning
- Identify basic cutting tools for CNC turning
- Describe the application of basic cutting tools for CNC turning
- Identify workholding devices for CNC turning
- Describe the application of workholding devices for CNC turning

Carefully read the unit, and then answer the following questions.

Multiple Choice

1. Which device is the most versatile for holding a variety of workpiece sizes?
- **a.** Collet
- **b.** Centers
- **c.** Chuck
- **d.** Live tooling

2. What type of machine typically uses a flat-bed design and is fitted with a carriage where tools are installed?
- **a.** Turret-type machine
- **b.** Gang tool machine
- **c.** Carousel tool machine
- **d.** Quick-change tool machine
- **e.** None of these choices

3. If machining a thin-walled, hollow workpiece or tubing, what would be the best workholding device?
- **a.** Collet
- **b.** Three-jaw chuck
- **c.** Four-jaw chuck
- **d.** Faceplate
- **e.** Power chuck

Name: ______________________ Date: ______________

4. To determine whether left-hand holders or right-hand holders are required for a turning application, what factor(s) need to be considered?
   a. Can the insert fit into the shape of the contour without clearance issues?
   b. Which direction will the tool be cutting, toward the headstock or away from the headstock?
   c. Will the tool be mounted in the machine upside down or right side up?
   d. Whether there is sufficient clearance between the toolholder and the workholding device.
   e. All of these choices are correct.
5. Swiss turning centers slide the stock in and out of the spindle for
   a. C-axis motion.
   b. X-axis motion.
   c. Y-axis motion.
   d. Z-axis motion.
6. Which of the following is not a collet chuck type?
   a. ER
   b. NM
   c. DA
   d. TG
7. A workholding collet that will consistently position the length of a part regardless of its diameter is called a
   a. Accu-set collet.
   b. Dead length collet.
   c. Fixed lock collet.
   d. Posi-set collet.
8. The device that threads onto the external threads of a collet chuck and constricts the collet when tightened is called a
   a. wedge nut.
   b. nose collar.
   c. end piece.
   d. tapered nut.
9. A grouping of several machines, all performing different operations on the same part is called a
   a. manufacturing cell.
   b. multi-flow group.
   c. multi-tasking line.
   d. efficiency co-op.

Name: ____________________ Date: ____________

True or False

10. Turning machines do not have an X-axis. ______
11. The spindle bore inside diameter limits the size of material used in collets. ______
12. Mounting and dismounting is performed by rotating a single screw on a VDI-type toolholder. ______
13. A CNC lathe has no ATC, but a turning center does. ______
14. A spindle collet can hold a wide range of workpiece diameters. ______
15. Live tools are motorized spindles that enable a turning center to perform milling and off-center hole-work. ______
16. A bar feeder is mounted to the machine turret and can pull the bar stock a desired distance for making the next part. ______

Short Answer

17. List three advantages of a gang tool holder machine configuration.

18. Briefly describe a VDI toolholder.

19. Describe the applications of grooving and turning tools and explain how they differ in geometry.

20. What is the purpose of a sub-spindle?

21. Describe two machining operations that can be performed by live tooling.

22. Briefly describe a split drill bushing and how it is used.

23. List three types of bar pullers.

24. List common types of cutting tools used in CNC turning machines.

Unit 3 CNC Turning: Programming

Name: ______________________________ Date: ______________

Score: ______________ Text pages 676–721

Learning Objectives

After completing this unit, you should be able to:

- Define basic G- and M-codes used for CNC turning
- Explain linear interpolation for CNC turning
- Explain circular interpolation for CNC turning
- Explain radial and diametral programming
- Describe facing operations for CNC turning
- Describe CNC drilling operations for CNC turning machines
- Describe tapping operations for CNC turning machines
- Explain the principles of tool nose radius compensation (TNRC) for CNC turning
- Describe CNC rough turning operations
- Describe CNC finish turning operations
- Describe boring operations for CNC turning machines
- Describe CNC grooving operations
- Describe threading operations for CNC turning machines
- Describe common canned cycles for CNC turning applications

Carefully read the unit, and then answer the following questions.

Multiple Choice

1. What code would be used for rapid, non-cutting axis movements?
   a. G01
   b. G19
   c. M0
   d. G00
   e. M05
2. When using cutter compensation, what code moves the tool to the right of the toolpath?
   a. G42
   b. G43
   c. G40
   d. G45
   e. None of these choices

Name: ______________________________ Date: ______________

3. What code is used for finishing canned cycles?
   a. G75
   b. G73
   c. G72
   d. G70
   e. G71
4. What types of part geometry are NOT affected by tool nose radius errors?
   a. Straight Z-motion
   b. Straight X-motion
   c. Tapers
   d. Radii
   e. Both "Straight Z-motion" and "Straight X-motion"
   f. Both "Tapers" and "Radii"
   g. All of the these choices
5. Which of the following best describes CNC finish turning operations?
   a. Drilling holes along the Z-axis using canned cycles
   b. Removing large amounts of material from the workpiece
   c. Producing desired surface finish and dimensional accuracy with light cuts
   d. None of these choices
6. The P-value of a Haas G82 drilling cycle specifies
   a. the number of seconds of dwell time.
   b. the number of micro seconds of dwell time.
   c. the number of milliseconds of dwell time.
   d. the number of revolutions of dwell.
7. What does the P-value specify in a G71 rough turning canned cycle?
   a. The first block sequence number defining the final part contour
   b. The last block sequence number defining the final part contour
   c. The depth of cut during roughing
   d. The amount of X-axis finish material
8. What does the U-value specify in a G71 rough turning canned cycle execution block (second line for Fanuc formats)?
   a. The first block sequence number defining the final part contour
   b. The last block sequence number defining the final part contour
   c. The depth of cut during roughing
   d. The amount of X-axis finish material
9. What is the angle for modified flank infeed threading?
   a. 15 degrees
   b. 29 degrees
   c. 30 degrees
   d. 60 degrees

Name: ______________________________ Date: ______________

10. What is the primary programming difference between OD G71 rough turning and ID G71 rough boring?
   a. The U-value is made negative.
   b. The R-value is made negative.
   c. No W-value is given.
   d. The X-value is made negative.

True or False

11. Code G00 is used for rapid movement. ________
12. Sequence numbers are entered behind a letter N. ________
13. The face of the workpiece is usually the Z-zero position of the part. ________
14. Facing cuts are typically performed from the inside of the workpiece toward the outer surface of the workpiece. ________
15. G01 is the code used to program an arc on a CNC lathe. ________
16. There are two ways to express feed rate units on a CNC lathe. ________
17. A G41 must be programmed to compensate for a tool's radius while cutting on the right of the part while in motion and a G42 when on the left. ________

Computations

18. Describe each variable of G75 X_ Z_ I_ K_ F_.

____

____

19. What is the feed rate for a 1/2-13 rigid tap operation? Round answer to four decimal places.

____

____

20. To perform a facing cut using a tool with a nose radius of 1/64, what is the programmed X-axis end-point (use four decimal places and use no spaces) at the center of the part?

____

____

21. Explain each function of the following codes: G0 G18 G20 G40 G54 G80 G99.

____

____

22. Explain each of the following codes: M4 S1200 T0202.

____

____

Name: ______________________ Date: ______________

Short Answer

23. List the three main motion types used in CNC programming.

____

____

24. What is TNRC and explain its use in CNC programming?

____

____

25. What is the common canned drill cycle code for CNC turning machines, and what letters are associated with the programming of the cycle?

____

____

26. What steps are taken at the beginning of a program to make a safe start and set default codes?

____

____

27. Briefly explain the similarity and difference between radial and diametral programming.

____

____

28. Briefly describe inside turning (boring) operations for CNC turning.

____

____

Unit 4 CNC Turning: Setup and Operation

Name: ______________________________ Date: ______________

Score: ______________ Text pages 722–736

Learning Objectives

After completing this unit, you should be able to:

- Explain CNC machine modes
- Explain the work coordinate system (WCS) for CNC turning
- Explain the machine coordinate system (MCS) for CNC turning
- Describe the homing procedure and purpose
- Explain workpiece offsets for CNC turning
- Explain tool geometry offsets for CNC turning
- Explain tool wear offsets for CNC turning
- Explain tool nose radius (or diameter) offsets
- Identify tool quadrant settings for TNRC
- Describe the three basic methods for loading programs into the MCU
- Explain program prove-out procedures

Carefully read the unit, and then answer the following questions in the space provided.

Multiple Choice

1. Using diameter programming, measurement reveals that each diameter on the first workpiece measures 0.0008″ larger than the desired size. What change should be made?
   - **a.** Add the 0.0008″ to the offset.
   - **b.** Subtract the 0.0008″ from the offset.
   - **c.** Add 0.0016″ to the offset.
   - **d.** Subtract 0.0004″ from the offset.
   - **e.** Add 0.0004″ to the offset.
2. In certain cases, instead of storing a program in the MCU memory, the program is fed from a PC to the control line by line as the machine runs the program. This is called
   - **a.** drip feeding.
   - **b.** manual data input.
   - **c.** direct numerical control.
   - **d.** USB download.
   - **e.** both "drip feeding" and "direct numerical control".

Name: ______________________________ Date: ______________

3. When mounting cutting tools in toolholders, check for
   a. clearance between cutting tool and holder.
   b. amount of cutting tool exposed.
   c. alignment of the cutting tool shank to the turret tool holder.
   d. tightness of cutting tool in holder.
   e. All of these choices are correct.
   f. None of these choices is correct.
4. The process of homing a CNC machine is to
   a. re-zero all axes.
   b. establish an MCS.
   c. recall the position of the WCS.
   d. establish travel limits.
   e. All of these choices are correct.
   f. None of these choices is correct.
5. The machine control mode used for entering and storing programs into memory is called
   a. zero-return mode.
   b. jog mode.
   c. MDI mode.
   d. edit mode.
   e. auto/memory mode.
6. The machine control mode used for manually moving the machine's axes to a desired position is called
   a. zero return mode.
   b. jog mode.
   c. MDI mode.
   d. edit mode.
   e. auto/memory mode.
7. The machine control mode used to run a stored program is called
   a. zero return mode.
   b. jog mode.
   c. MDI mode.
   d. edit mode.
   e. auto/memory mode.
8. The button that begins the active CNC program and is colored green on many machines is called
   a. go.
   b. execute.
   c. cycle start.
   d. run.
9. Why is it important to enter the correct tool tip quadrant orientation when using tool nose radius compensation (TNRC)?
   a. To ensure the workpiece offset is properly activated during TNRC
   b. To define the area of the workpiece where machining occurs
   c. To allow the machine to appropriately compensate for the tool nose radius when using TNRC
   d. None of these choices

Name: ______________________ Date: ____________

True or False

10. DNC can be used to run programs too large to fit in the MCU. ______

11. Programs entered using MDI are stored in program memory. ______

12. The work coordinate system identifies the location of the origin on the workpiece. ______

13. The tool approach quadrant must be entered in the geometry offset page when using tool nose radius compensation. ______

14. The homing process recalls the MCS position. ______

15. The MCS can change with each program. ______

Computations

16. Using diameter programming mode and the method described in this unit, if a tool is touched off on a 1.250″ diameter with a 0.015″ feeler gage, what is the value that will be entered into the geometry offset page?

____

____

17. Using the method described in this unit, if a tool is touched off on the Z-zero surface of the part with a 0.015″ feeler gage, what is the value that will be entered into the geometry offset page?

____

____

Short Answer

18. Describe the five modes in which a CNC machine can be operated?

____

____

19. Explain the term *work offset.*

____

____

20. What control feature can be used to adjust for tool wear during a production run?

____

____

21. Explain the process of aligning a hole working tool with the center of the workpiece.

____

____

Name: ____________________ Date: ____________

22. Explain three ways to enter a program in the MCU.

____

____

23. What are three ways to prove-out a program?

____

____

24. Explain the importance of the tool nose radius offset.

____

____

Unit 5 Introduction to CNC Milling

Name: ______________________________ Date: ______________

Score: ______________ Text pages 737–751

Learning Objectives

After completing this unit, you should be able to:

- Describe different types of CNC milling machines
- Explain the machine axes used for milling
- Describe the two major types of ATCs
- Describe the uses of workholding devices for CNC milling
- Describe uses of toolholding devices used for CNC milling

Carefully read the unit, and then answer the following questions.

Multiple Choice

1. On a CNC machining center, tools are held in the spindle by the
  a. taper.
  b. toolholder.
  c. retention knob.
  d. arbor.
  e. flange.

2. A BT-40 holder uses a(n)
  a. inch series threaded retention knob.
  b. metric threaded retention knob.
  c. thinner flange than a CAT-40 holder.
  d. groove in the center of the flange.
  e. R-8 taper.

3. Select the toolholder for endmills that has the least run-out.
  a. Solid toolholder with set screw
  b. Floating holder
  c. Collet chuck
  d. Arbor chuck
  e. Drill chuck holder

4. The axes of a horizontal machining center (HMC) can be described as
  a. X and Z axes only with the spindle aligned to the Z-axis.
  b. X, Y, and Z axes, with the spindle oriented horizontally.
  c. a polar coordinate system.
  d. X, Y, and Z axes, with the spindle oriented vertically.
  e. None of these choices is correct.

Name: ______________________________ Date: ______________

5. What device is gripped by a drawbar mechanism in the machine spindle to secure toolholders?
   a. Retention knob
   b. Grab bar
   c. Collet closer
   d. Ball gripper
6. Most machining center spindles use an NMTB taper size, which has a taper of
   a. 10 degrees included.
   b. 3 ½″ per foot.
   c. 7″ per foot.
   d. 5 degrees included.
7. The combination of all machining operations required to produce a part is called a manufacturing
   a. sequence.
   b. operation.
   c. process.
   d. procedure.
8. A clamp that can secure and release a part by flipping a hand-operated lever is called a
   a. cam-action clamp.
   b. toggle clamp.
   c. tool-less clamp.
   d. toe clamp.

True or False

9. CNC machining centers are categorized by their knee size. ________
10. The swing-arm tool changer is faster than the carousel changer. ________
11. One difference between a CAT and BT toolholders is the flange thickness. ________
12. Some CNC milling machines do not use either CAT or BT flange holders. ________
13. Taps used in a machining center must always be held with floating holders. ________
14. The two types of automatic tool changers found in machining centers are the carousel-type and the Ferris wheel-type. ________
15. Machining centers can be separated into two main types: vertical and horizontal spindle. ________
16. Shrink-fit toolholders accept TG-style collets. ________

Name: ______________________________ Date: ______________

Identification Matching

17. Select the best workholding device for the given machining operations.

a. Clamps	________	**1.** Used to hold thin, smooth, flexible parts flat for light machining
b. Vises	________	**2.** Used to hold and locate cylindrical parts
c. Tombstones	________	**3.** Used to hold extremely large or oddly shaped workpieces
d. Vacuum plate	________	**4.** Device specifically designed to accommodate a specific part
e. Custom fixture	________	**5.** Used to hold a solid part with parallel edges
f. Collet fixture	________	**6.** A device that uses two or more workholding tooling plates mounted flat on the machine table
g. Pallet system	________	**7.** Used to mount multiple workholding devices at one time. The device must rotate to access all the mounted workpieces

Short Answer

18. What are two advantages of using a shrink-fit toolholder?

____

____

19. Name the two most commonly used toolholder flange types.

____

____

20. List five machining operations that a milling-type CNC can perform.

____

____

21. Explain the procedure for securing an endmill in an endmill holder.

____

____

22. List three types of collets often used to hold tools in machining centers.

____

____

Section 8: Computer Numerical Control

Unit 6 CNC Milling: Programming

Name: ______________________________ Date: ______________

Score: ______________ Text pages 752–793

Learning Objectives

After completing this unit, you should be able to:

- Identify basic G- and M-codes used for CNC milling
- Explain linear interpolation for CNC milling
- Explain circular interpolation for CNC milling
- Explain the arc center method for circular interpolation
- Explain the radius method for circular interpolation
- Describe facing operations for CNC milling
- Describe two-dimensional CNC milling
- Describe drilling and tapping canned cycles for milling
- Describe cutter radius compensation for milling

Carefully read the unit, and then answer the following questions.

Multiple Choice

1. The code for feed rates expressed in inches per minute (IPM) is
 a. G17.
 b. G90.
 c. G94.
 d. G95.
 e. G93.

2. Each block of code ends with an End of Block character, also called a
 a. semicolon.
 b. colon.
 c. parentheses.
 d. sequence number.
 e. sequence space.

3. In CNC mill programming, it is a standard practice to establish a clearance plane above the workpiece or workholding device. Choose the best clearance plane from the choices below:
 a. .625″
 b. .010″
 c. 0.100″
 d. 0.500″
 e. All these choices

Name: ______________________________ Date: ______________

4. The code used to stop all programmed functions during a machining operation is a
   a. M01.
   b. M00.
   c. M03.
   d. M05.
   e. M09.
5. Which letter is associated with the tool height offset?
   a. I
   b. J
   c. K
   d. H
   e. T
6. Work Coordinate System #3 is activated with which code?
   a. G56
   b. G90
   c. G21
   d. G30
7. The arc center method of circular interpolation requires the additional letters I and J. What do these letters define?
   a. The distance from the arc start position to the arc end position
   b. The distance from the arc center position to the arc start position
   c. The distance from the arc start position to the arc center position
   d. The distance from the arc center position to the arc end position
8. Which code commands the tool to retract to the initial plane between hole locations during a canned drilling cycle?
   a. G81
   b. G83
   c. G98
   d. G99
9. What must be done when using the R-method to program an arc that is greater than 180 degrees but less than 360 degrees?
   a. A G98 code must be used.
   b. A negative R-value must be used.
   c. A negative J-value must be used.
   d. A positive I-value must be used.
10. What code sets the XY plane during the safe-start portion of a program?
   a. G17
   b. G18
   c. G20
   d. G91

Name: ____________________ Date: ____________

True or False

11. When rigid tapping, the tap must be held in a floating holder. ______
12. The values of the H-word and the T-word should always match. ______
13. G41 cutter comp is used for conventional milling. ______
14. G74 is used for right-hand tapping in canned cycles. ______
15. A D-word is used to activate a tool radius offset value that is stored in the MCU. ______
16. The return plane for a canned drilling cycle can be different from the initial plane. ______
17. The movement used to activate and cancel automatic cutter radius compensation must be a G0 motion. ______

Identification Matching

18. Match the proper code to the given command.

a. G95	______	**1.** Spindle off
b. G90	______	**2.** Coolant on
c. G20	______	**3.** Spindle on clockwise
d. M05	______	**4.** Inches per revolution feedrate selection
e. M08	______	**5.** Absolute positioning
f. M03	______	**6.** Inch unit selection

Short Answer

19. What command would be given to start the spindle in a counterclockwise direction at 250 RPM?

20. What command would be given to execute a tool change to tool #8?

21. Write a block of code using the radius method to machine a clockwise 0.2″ radius ending at the position X1.2, Y1.2. The tool is already at the start position.

22. Write a block of code to activate tool height offset compensation and bring tool #5 within 0.1″ of the Z-zero surface.

23. List codes for, and explain the differences between single-pass, fast-peck, and deep-hole drilling canned cycles.

Name: ______________________________ Date: ______________

24. Explain each function of the coding: G0 G54 X5.5 Y3.0.

____

____

25. Explain linear interpolation and the command used for CNC milling.

____

____

26. Explain circular interpolation and the command used for CNC milling.

____

____

27. What is the purpose of the facing operation in CNC milling, and when is it typically performed?

____

____

28. What is two-dimensional milling in CNC, and which operations are performed using this method?

____

____

Unit 7 CNC Milling: Setup and Operation

Name: ______________________ Date: ______________

Score: ______________ Text pages 794–814

Learning Objectives

After completing this unit, you should be able to:

- Describe CNC machine modes for CNC milling
- Describe the work coordinate system (WCS) for CNC milling
- Describe the machine coordinate system (MCS) for CNC milling
- Describe the homing procedure and purpose for CNC milling
- Describe workpiece offsets for CNC milling
- Describe tool geometry offsets for CNC milling
- Describe tool wear offsets for CNC milling
- Describe cutter radius compensation offsets
- Describe the three basic methods for loading programs into the MCU
- Explain program prove-out procedures for CNC milling

Carefully read the unit, and then answer the following questions.

Multiple Choice

1. What machine mode is selected to manually move a machine axis?

a. MDI
b. Auto
c. Jog
d. Edit
e. Zero-reference

2. What machine mode should be selected to enter and save a program into the MCU?

a. MDI
b. Auto
c. Jog
d. Edit
e. Zero-reference

3. In an emergency or collision with a(n) object, what button should be pushed?

a. Feed hold
b. Feed override
c. Rapid override
d. Red emergency stop button
e. All of these choices

Name: ______________________________ Date: ______________

4. Initial tool height offsets are stored in the
   a. work offset page.
   b. geometry offset page.
   c. wear offset page.
   d. MCS offset page.
5. A tool height offset adjustment used to compensate for a failing tool are stored in the
   a. work offset page.
   b. geometry offset page.
   c. wear offset page.
   d. MCS offset page.
6. During set-up, what would be the best tool to accurately locate a workpiece that has a hole in the center?
   a. Dial indicator
   b. Edge finder
   c. Wiggler
   d. Master ball
7. The Z-axis part origin is established with which offset?
   a. Tool height offset
   b. Work offset
   c. Geometry offset
   d. Wear offset

True or False

8. The homing process is used to recall the MCS and WCS of the machine. ________
9. The length of a tool is measured from the gage line on the toolholder's taper to the tool tip. ________
10. The tool radius value entered in the offset register is ignored unless cutter comp is activated. ________
11. Wear offset adjustments can be made to either the tool height or the tool radius offset. ________
12. A positive wear compensation value adjustment will produce smaller external dimensions. ________
13. The work coordinate system is moveable, and the origin can be placed anywhere on the workpiece. ________
14. The machine coordinate system is moveable, and its origin can be changed or moved. ________

Name: ______________________ Date: ______________

Computations

15. Using the method described in this unit, if the spindle face is touched off of the Z-zero surface of the part with a 2.0000 gage block, and the machine position register says the position is –18.5678, what is the total work offset for the Z-axis? (Include proper sign.)

____

____

16. Using the method described in this unit, if a tool is touched off the Z-zero surface with a 0.015 feeler gage and the absolute (work) position register says the position is 4.5432, what is the total length of the tool?

____

____

17. Assume that a tool radius value of 0.2500 is entered initially in the tool geometry offset page. The wear offset is set to zero. Using cutter radius compensation, 200 parts have been made within tolerance. Then the tool begins producing parts that are too large in their overall length and width dimensions by a total of 0.0015. What value must be entered in the wear offset register to correct the problem?

____

____

Short Answer

18. Explain one way to run a machine program that is too large to fit in the memory of the control.

____

____

19. According to the methods described in the book, which must be set first, the Z work offset or the tool height offset? Why?

____

____

20. If a part feature is cut too deep because of a tool offset error, should the offset be shortened or lengthened? Why?

____

____

Name: ______________________________ Date: ______________

21. When offset adjustments are necessary, explain when it is appropriate to use a geometry offset and when it is appropriate to use a wear offset.

____

____

22. What are three basic methods for loading programs into the Machine Control Unit (MCU).

____

____

23. List four methods used to verify a CNC milling program before allowing the machine to run unsupervised.

____

____

Section 8: Computer Numerical Control

Unit 8 Computer-Aided Design and Computer-Aided Manufacturing

Name: ____________________ Date: ____________

Score: ____________ Text pages 815–826

Learning Objectives

After completing this unit, you should be able to:

- Describe the basic applications of CAD
- Describe the basic applications of CAM
- Describe wireframe drawings
- Describe solid model drawings
- Describe surface drawings
- Describe the basic principles of toolpath creation
- Describe basic toolpath types
- Describe the basic principles of post-processing

Carefully read the unit, and then answer the following questions.

Multiple Choice

1. Which endmill is best for 3-D contour milling?
   - **a.** Center-cutting endmill
   - **b.** Four-flute flat endmill
   - **c.** Two-flute flat endmill
   - **d.** Ball mill
   - **e.** Shell endmill

2. What part of the CNC programming process involves the creation of a CNC program code?
   - **a.** Toolpath creation
   - **b.** Geometry creation
   - **c.** Post-processing
   - **d.** Wire framing
   - **e.** Solid modeling

3. To improve the surface roughness remaining from the cusp height, the user may also desire to switch to a different
   - **a.** pattern direction of the tool during finishing.
   - **b.** change of cutting tools.
   - **c.** change of speed and feeds.
   - **d.** adjustment of tool offsets.
   - **e.** adjustment of tool height offsets.

Name: ______________________________ Date: ______________

4. A connected group of selected entities is called a
   a. toolpath.
   b. chain.
   c. model.
   d. contour.
5. The ridges left from a ball mill during a surfacing operation are called
   a. cusps.
   b. step-over peaks.
   c. hatch grooves.
   d. ridgules.
6. What drawing type is two-dimensional and outlines the basic part shape?
   a. Dynamic model
   b. Solid model
   c. Wireframe
   d. Surface model
7. What drawing type appears three-dimensional and as though it has mass?
   a. Dynamic model
   b. Solid model
   c. Wireframe
   d. Surface model

True or False

8. Turning operations are often programmed longhand without using CAM software. ________
9. CAD/CAM eliminated the use of paper to convey the part drawing. ________
10. CAD software uses post-processing operations. ________
11. Post-processors write code generically and must be edited afterward for specific machine control formats. ________
12. Boundaries are set to limit tool travel in 3D surface milling. ________
13. Not all CAM software packages require the part to be drawn in a separate CAD software system. ________
14. In order to create toolpaths, connected entities are first selected in the order of machining operation. ________

Name: ______________________ Date: ______________

Short Answer

15. What are the three major steps to creating a CNC program using CAM software?

____

____

16. List the three types of CAD drawing forms.

____

____

17. What do the letters in CAD and CAM stand for?

____

____

18. List four toolpath examples for milling operations.

____

____

Drill Drift

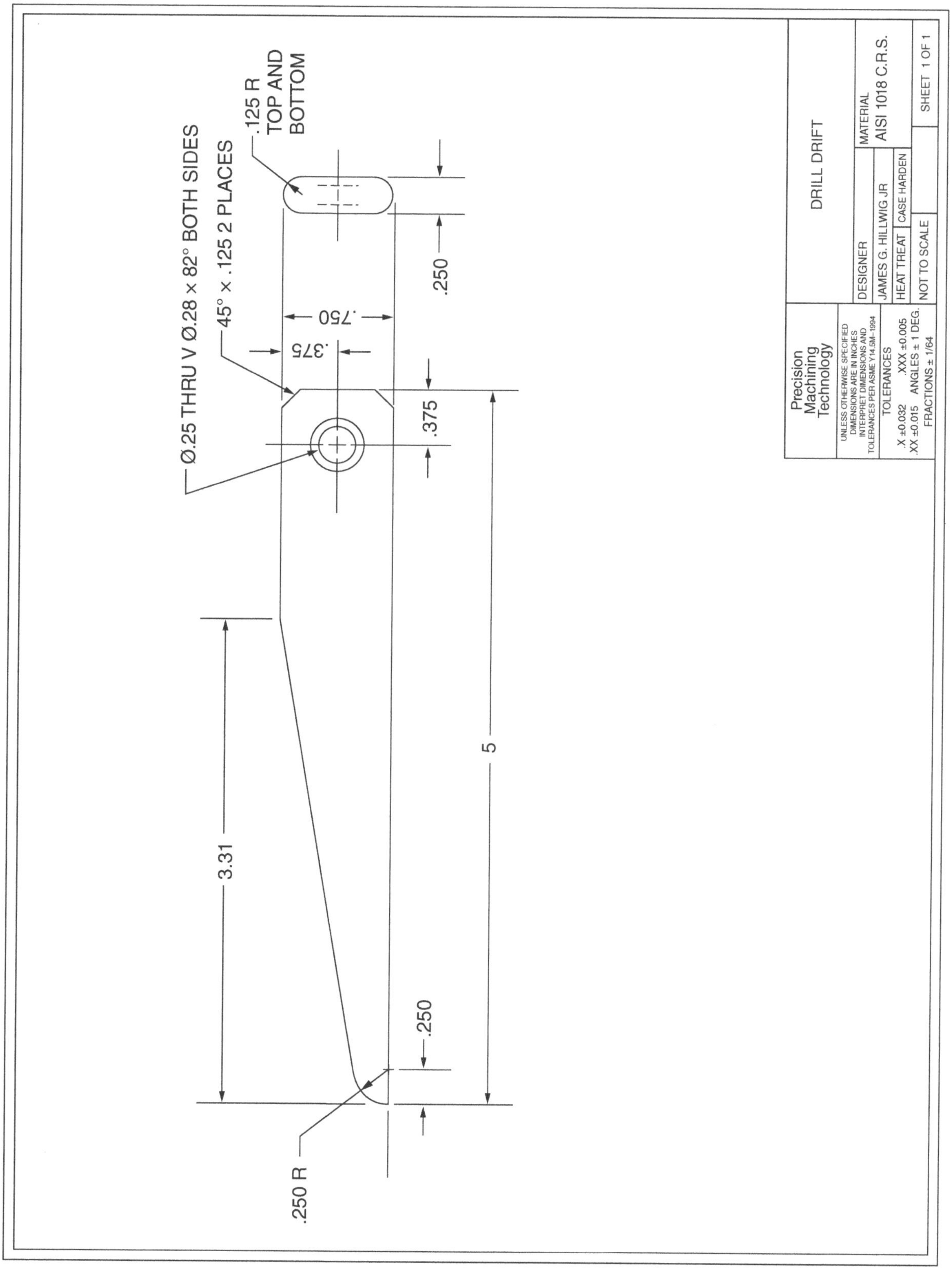

Ø.25 THRU V Ø.28 × 82° BOTH SIDES
45° × .125 2 PLACES
.125 R
TOP AND
BOTTOM
.375
.750
.250
.375
3.31
5
.250
.250 R
DRILL DRIFT
Precision
Machining
Technology
UNLESS OTHERWISE SPECIFIED
DIMENSIONS ARE IN INCHES
INTERPRET DIMENSIONS AND
TOLERANCES PER ASME Y14.5M-1994
TOLERANCES
.X ±0.032 .XXX ±0.005
.XX ±0.015 ANGLES ± 1 DEG.
FRACTIONS ± 1/64
DESIGNER
JAMES G. HILLWIG JR
MATERIAL
AISI 1018 C.R.S.
HEAT TREAT
CASE HARDEN
NOT TO SCALE
SHEET 1 OF 1

NIMS Skills Practiced

- Job Process Planning
- Layout
- Benchwork
- Drill Press

Equipment

Power saw, bench vise, drill press

Tools Needed

3″ to 6″ dividers
Layout dye
Hammer
File card
82° countersink
Scribe
Prick punch
Hacksaw
Abrasive cloths
6″ steel rule
Center punch
File
1/4″ drill bit

Material Needed

- 1/4″ × 3/4″ × 5-1/8″ mild steel

Safety

1. Follow all shop safety rules given to you by your instructor and be sure to pass any safety tests given before using shop equipment.
2. Follow all safety rules for power equipment.
3. Be careful with sharp tools when using them.
4. Make sure the file has a handle on it.

Order of Operations

1. Look over blueprint to determine tolerances, tools, and materials needed.
2. Gather tools needed, figure out RPMs for required tools, and write down their RPMs on sheet.
3. Measure, mark, and cut stock 1/8″ longer than length on print so you can square up one end.
4. Square one end and deburr for layout.
5. Paint with dye and lay out part.
6. Drill 1/4″-diameter hole, then countersink both sides with 82° countersink.
7. Cut with hacksaw close to your layout lines, staying 1/32″ away.
8. Finish by filing the outside of part to layout lines, being careful to follow print dimensions.
9. File radii where indicated on print.
10. Polish with abrasive cloths.
11. Inspect the part to print dimensions; record on inspection sheet and turn in.

How to Determine the RPMs

Tool Being Used	Surface Feet per Minute	Formula	RPM
1/4" drill bit	80	$\frac{4 \times 80}{.250}$	
82° countersink	80	$\frac{4 \times 80}{\text{Diameter of Countersink}}$	

Student Name: ______________________ Date Submitted: __________

Class: ______________ Total Hours on Job: __________

	Print Dimensions and Tolerances	Student's Inspection	Instructor's Inspection	Instructor Comments
1	**{Length}** 5 Long ± .015			
2	**{Width}** .250 Wide ± .005			
3	**{Height}** .750 High ± .005			
4	**{Taper}** 3.31 Long ± .015			
5	**{Front Radius}** .250 R ± .005			
6	**{Height to Front Radius}** .250 R ± .005			
7	**{Distance from End to Radius}** .250 ± .005			
8	**{Rear Radius}** .125 × 45° 2 places			
9	**{Hole}** Ø .25 ± .015			
10	**{82° Countersink}** Ø .25 ± .015			
11	**{Hole Ø .25 Location}** .375 × .375 ± .005			
12	**{Top Radius}** .125 ± .005			
13	**{Bottom Radius}** .125 ± .005			
14	Is the part free of burrs and does it meet finish requirements?			
15	Is the part properly identified?			

Project Grade: __________

"T" Slot Mill Cleaner

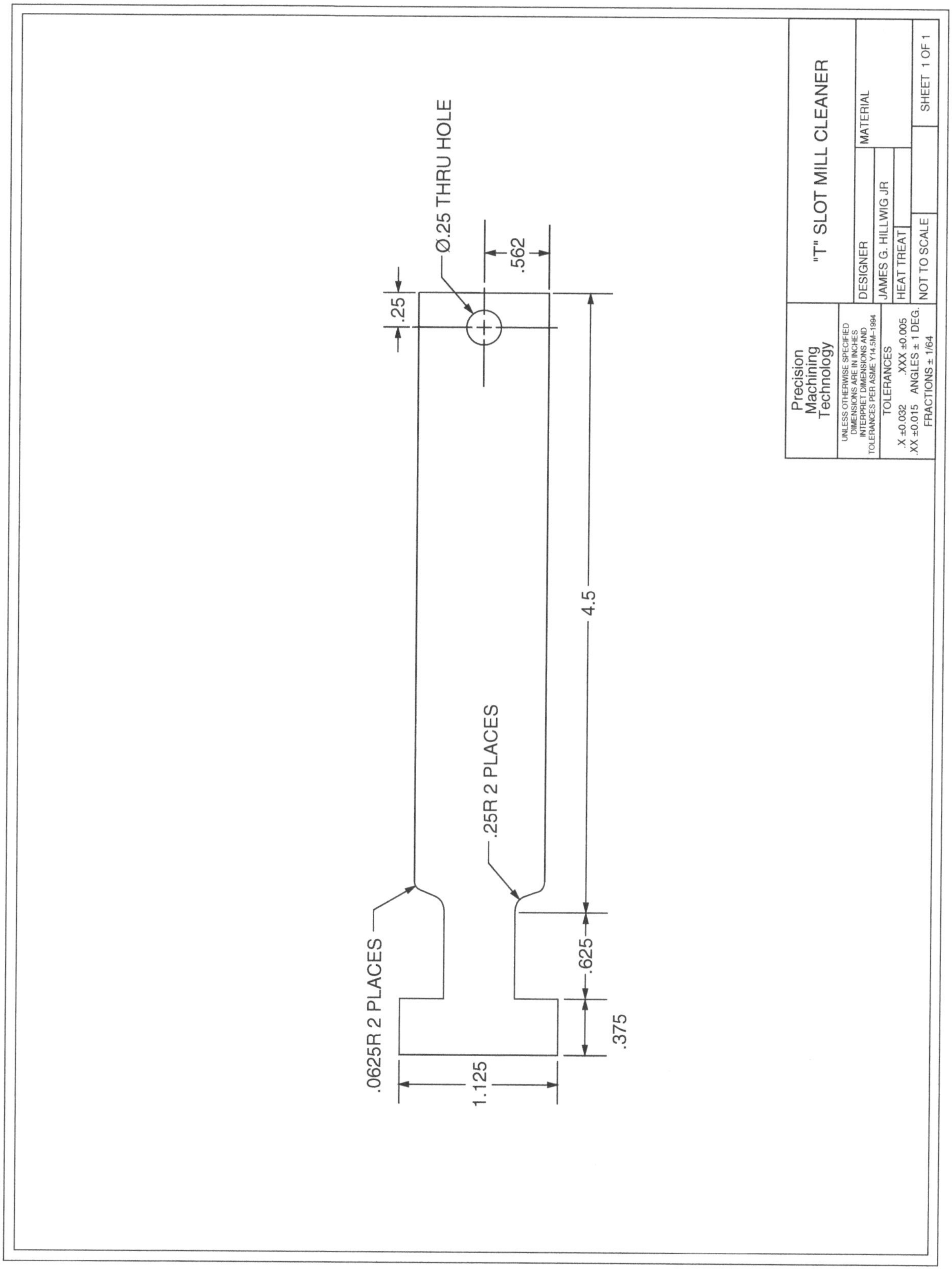

NIMS Skills Practiced

- Job Process Planning
- Layout
- Benchwork

Equipment

Power saw, bench vise, drill press

Tools Needed

3″ to 6″ dividers	Scribe	12″ combination set
Layout dye	Prick punch	Center punch
Hammer	Hacksaw	File
File card	Abrasive cloths	Radius gages
Center drill	1/4″ drill	Countersink

Material Needed

- 1/8″ × 1-3/16″ × 4-1/2″ aluminum or CRS

Safety

1. Follow all shop safety rules given to you by your instructor and be sure to pass any safety tests given before using shop equipment.
2. Follow all safety rules for power equipment.
3. Be careful of sharp tools when using them.
4. Make sure the file has a handle on it.

Order of Operations

1. Look over blueprint to determine tolerances, tools, and materials needed.
2. Gather tools needed.
3. Measure, mark, and cut stock 1/8″ longer than length on print so you can square up one end.
4. Using file and square, square up one end and deburr for layout.
5. Paint with dye and lay out part.
6. Cut with hacksaw close to your layout lines, staying 1/32″ away.
7. Finish by filing the outside of part to layout lines, being careful to follow print dimensions.
8. File radii where indicated on print.
9. Center drill, drill, and countersink 1/4″ hole.
10. Polish with abrasive cloths.
11. Inspect the part to print dimensions; record on inspection sheet and turn in.

Student Name: ______________________ Date Submitted: ____________

Class: ______________ Total Hours on Job: __________

	Print Dimensions and Tolerances	Student's Inspection	Instructor's Inspection	Instructor Comments
1	Is the part free of burrs and does it meet finish requirements?			
2	Is the part properly identified?			
3	**{Length}** 5.5 ± .032			
4	**{Height}** 1.125 ± .005			
5	**{Height}** 1.000 ± .005			
6	.375 ± .005			
7	**{Neck}** .500 ± .005			
8	.0625 R ± .005 6 places			
9	.25 R 2 places ± .015			
10	.625 × .500 Neck ± .005			
11	Ø .25 ± .015 Thru Hole			
12				
13				
14				
15				
16				
17				
18				
19				
20				

Project Grade: ______________

Soft Jaws for Bench Vise

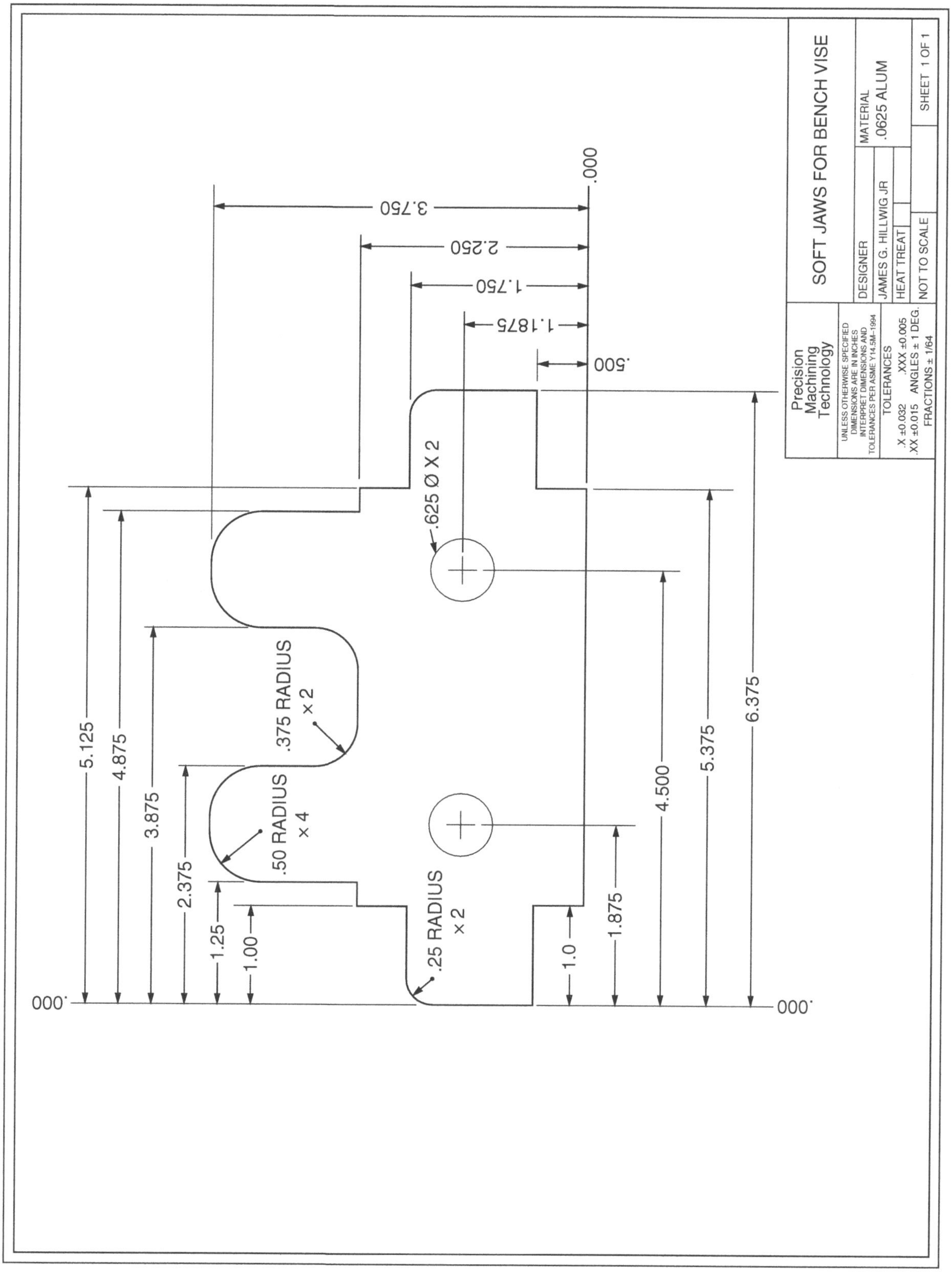

NIMS Skills Practiced

- Job Process Planning
- Layout
- Benchwork

Equipment

Power saw, bench vise

Tools Needed

3″ to 6″ dividers	Scribe	12″ combination set
Layout dye	Prick punch	Center punch
Hammer	Hacksaw	File
File card	Abrasive cloths	Radius gages

Material Needed

- 1/16″ × 4″ × 6-1/2″ aluminum

Safety

1. Follow all shop safety rules given to you by your instructor and be sure to pass any safety tests given before using shop equipment.
2. Follow all safety rules for power equipment.
3. Be careful of sharp tools when using them.
4. Make sure the file has a handle on it.

Order of Operations

1. Look over blueprint to determine tolerances, tools, and material needed.
2. Gather tools needed.
3. Measure, mark, and cut stock 1/8″ longer than length on print so you can square up one end.
4. Using file and square, square up one end and deburr for layout.
5. Paint with dye and lay out part.
6. Cut with hacksaw close to your layout lines, staying 1/32″ away.
7. Finish by filing the outside of part to layout lines, being careful to follow print dimensions.
8. File radii where indicated on print.
9. Polish with abrasive cloths.
10. Inspect the part to print dimensions; record on inspection sheet and turn in.

Student Name: ______________________ Date Submitted: ____________

Class: ____________ Total Hours on Job: ____________

	Print Dimensions and Tolerances	Student's Inspection	Instructor's Inspection	Instructor Comments
1	Is the part free of burrs and does it meet finish requirements?			
2	Is the part properly identified?			
3	**{Length}** 6.375"			
4	**{Width}** 3-3/4"			
5	**{Holes}** 5/8" Diameter			
6	2-5/8" Apart			
7	**{Bottom}** 4.375" Length			
8	1" Length (2 places)			
9	.50" Height (2 places)			
10	**{Sides}** 1.25" Height			
11	.25" R (2 places)			
12	.50" Height (2 places)			
13	1" Length (2 places)			
14	1.500" Height (4 places)			
15	.250" Length (2 places)			
16	.50" Height (2 places)			
17	**{Top}** .5" R (4 places)			
18	1-1/8" (2 places)			
19	3/8" R (2 places)			
20	1-1/2" Length			

Project Grade: ____________

C-Clamp

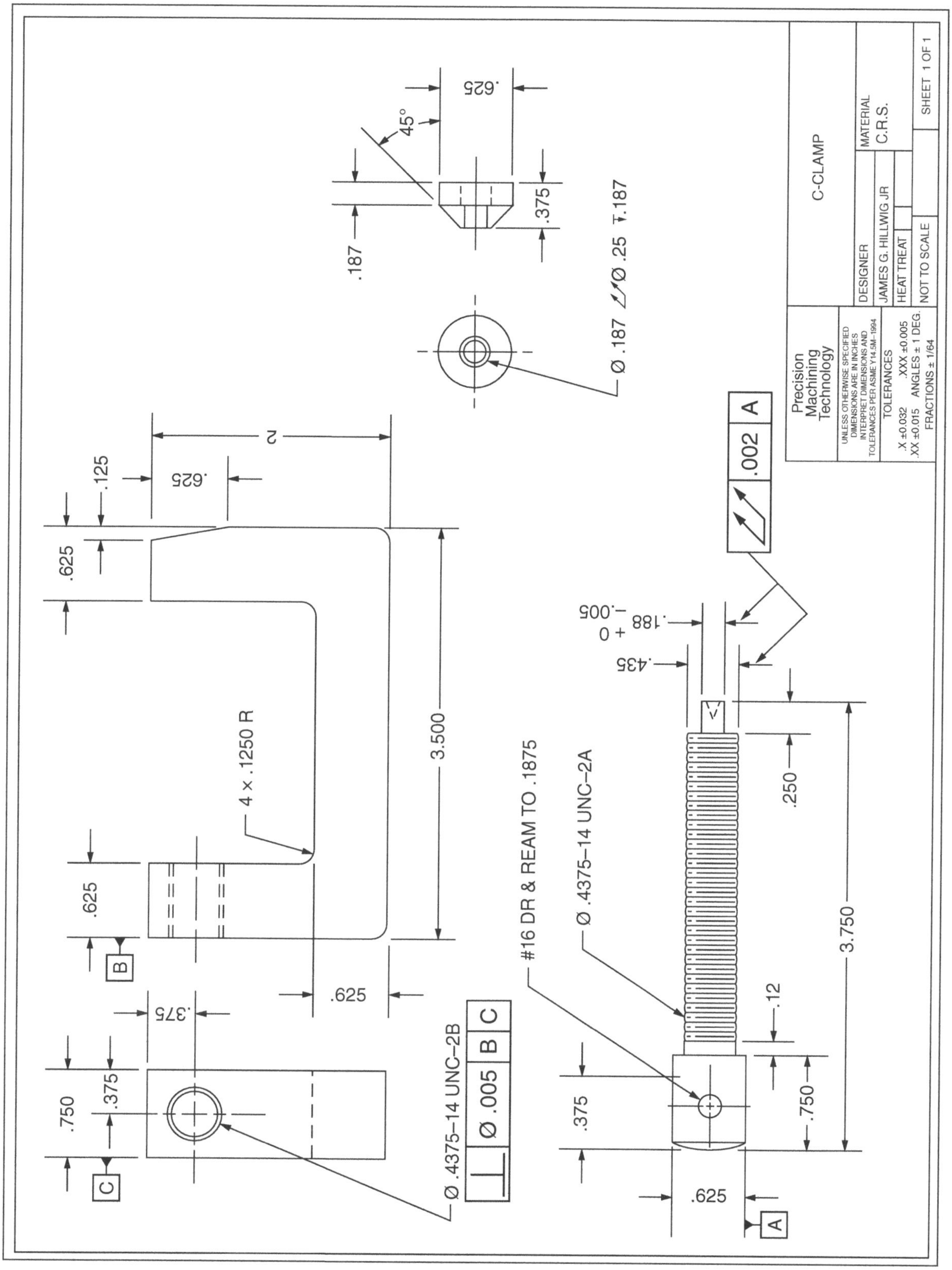
C-CLAMP
DESIGNER
JAMES G. HILLWIG JR
MATERIAL
C.R.S.
HEAT TREAT
NOT TO SCALE
SHEET 1 OF 1
Precision Machining Technology
UNLESS OTHERWISE SPECIFIED DIMENSIONS ARE IN INCHES INTERPRET DIMENSIONS AND TOLERANCES PER ASME Y14.5M-1994
TOLERANCES
.X ±0.032 .XXX ±0.005
.XX ±0.015 ANGLES ± 1 DEG.
FRACTIONS ± 1/64
4 × .1250 R
3.500
#16 DR & REAM TO .1875
Ø .4375–14 UNC–2A
Ø .4375–14 UNC–2B
Ø .005 B C
.002 A
Ø .187 Ø .25 .187
45°
.188 + 0 −.005
.435
3.750

NIMS Skills Practiced

- Job Process Planning
- Layout
- Turning Operations: Chucking
- Drilling and Tapping
- Surface Grinding and Safety
- Vertical Milling and Safety
- Squaring up a Block

Equipment

Power saw, milling machine, lathe, surface grinder

Tools Needed

Scribe
Dead blow hammer
Various drill bits
90° countersink
Edge finder
Machinist square
Tramming items for mill
Various types of turning, threading, and facing tools
6″ steel rule
File
7/16-14 tap
Various sizes of endmills
Center drill
Grinding wheel
Adjustable mill angle
3/16″ reamer
Layout dye
File card
Stamps
Parallels
Vise stop
Grinder vise
Thread triangles
Magnetic sine plate

Material Needed

- 3/4″ thick × 2″ wide × 3-5/8″ long CRS (for "C" body)
- 5/8″ diameter × 4″ long CRS (for screw)
- 5/8″ diameter bar about 12″ long (to safely hold cap while machining)
- 3/16″ diameter × 1-1/2″ long CRS (for screw handle)

Safety

1. Follow all shop safety rules given to you by your instructor and be sure to pass any safety tests given before using shop equipment.
2. Follow all safety rules for power equipment.
3. Be careful of sharp tools when using them.
4. Make sure the file has a handle on it.
5. Be sure all cutting tools are sharp and care is taken while handling them.
6. Be sure ways are locked on mill when necessary.
7. Be sure chuck is tightened and chuck wrench is never left in chuck.
8. Make sure stock is supported with center when necessary.
9. Ring grinding wheels before using them.
10. Make sure magnetic chuck is on when surface grinding.

Order of Operations

MILL

1. Look over blueprint to determine tolerances, tools, and material needed.
2. Gather tools needed, figure out RPMs for required tools, and write down their RPMs on sheet.
3. Measure, mark, and cut stock 3/4" thick × 2" wide. Saw 3-5/8" long.
4. Make sure milling machine is in tram.
5. Deburr all over.
6. Place piece in mill vise on parallels with sawed end sticking out; tap down with dead blow hammer to seat part on parallels, then mill just enough off the end to clean it up.
7. Deburr the end you just milled, flip part 180° on parallels in vise, tap down with dead blow hammer to seat part on parallels, and mill to length.
8. Lay out part.
9. Place part in vise on parallels against vise stop, tap down with dead blow hammer to seat part on parallels, and set vise stop on left side of part.
10. Insert edge finder in collet, set RPMs to 1100, and edge find the X-axis and the Y-axis. Be sure to allow for 1/2 grind stock on the X-axis.
    a. X-axis should read −.1075 (this includes 1/2 of the edge finder and 1/2 of the grind stock).
    b. Y-axis should read .1000 (with the tolerance on the print, the 2" stock size should leave plenty of room for grinding).
11. Place center drill in chuck, and move in 5/8" + .015 + 1/8" on the X-axis.
12. Take 2"—(5/8" + .015 + 1/8"); this will be your Y-axis location.
13. Lower center drill to make sure it aligns with layout lines.
14. If location is correct, center drill hole at 1200 RPMs, change to 1/4" drill, set the RPMs, and drill through part.
15. Remove part and deburr, flip part end for end against your stop and repeat the center drill and drilling operations, then remove part and deburr. If done correctly, the holes should be perfect.
16. Now stand part up on parallels, holding the bottom of the "C" in the vise with holes you drilled just above the top of the vise jaws.
17. Place a 3/4" roughing endmill in a collet, keeping away from the 5/8" layout line on the side, and mill down until you reach the center of the hole.
18. Using a band saw, you can now cut to the center of the remaining hole from the top and slide the blade in the 3/4" slot you just cut along the bottom of the "C" to meet where you just cut. This will remove the bulk of the center.
19. Using a finishing endmill, place the part back in the vise and clean up the sides and bottom, leaving grind stock. DO NOT REMOVE THE RADIUS FROM THE CORNERS.
20. Now stand up on end against the vise stop, keeping the 2" side along vise jaws, and move to the 7/16-14 hole location.
21. Center drill, and drill with a letter "U" drill, countersink the hole, then power tap the hole with a 7/16-14 tap.
22. Trig out the angle; with the 1/8" and 5/8" locations of the angle laid out, set the opposite corner of the drilled end of the part on the mill angle, and mill the angle to layout lines. DO NOT remove lines; just touch them—this will leave grind stock. (Have instructor check setup before you begin your cut.)
23. Deburr part and hold with the bottom of the "C" up against the vise stop. Using a 3/8" endmill, cut a name pocket centered both ways .030 deep, 3/8" wide × 3/4" long.
24. Now hold in a bench vise and file the 1/8" radius on the bottom corners of the part.
25. Deburr, place on anvil, and using steel stamps, stamp your name in pocket.

*** This part can be heat treated if your budget permits, but it is not necessary.***

SURFACE GRINDER

MAKE SURE MAGNETIC CHUCK IS ON WHEN SURFACE GRINDING!

26. Surface grind all sides square and parallel. Deburr after each side is ground.

27. Surface grind the face to clean it up, deburr, then flip the part to grind the opposite side.

28. Deburr all edges and place all the way down in grinder vise with ground sides against jaws, leaving one edge sticking out of side of vise so you can turn vise on its side.

29. Grind top of part to clean up and turn vise on side without removing or disturbing part while in vise and grind just enough to clean it up.

30. Remove from vise, deburr all edges, place ground side down on chuck with square blocking on each side, and grind bottom to clean it up.

31. Deburr and place last side up with square blocking on each side and clean up surface.

32. Now check sizes and hole location; grind them into location and part to size. Be sure to deburr and block part when needed.

33. Using vise and magnetic sine bar, calculate gage block height, set magnetic sine bar to height, place part in grinder vise, place on magnetic sine bar, TURN ON MAGNETS, and grind angle.

34. Using a sine bar and roll dimensions, check the angle. (Have your instructor help you.)

35. Dressing a radius on your grinding wheel: grind the inside radius on the clamp. (See instructor.)

36. Lay flat on grinder chuck and grind a .030 × 45° chamfer on all edges.

LATHE

37. Chuck 5/8" diameter × 4" long stock in lathe face until end is clean.

38. Mark faced end at .750; this will be the head of the screw.

39. File the convex on the face of the screw.

40. Turn part end for end, face to length, and center drill 3/16" diameter using a #3 center drill. The center-drilled hole remains in the part this time.

41. Turn .435 diameter back 3" to the .750 mark of the head.

42. Chase the thread on the screw.

43. Turn the .188 diameter back 1/4".

44. Set screw in mill or drill press and drill and ream the 3/16" hole in the head.

45. Chuck the 5/8" diameter bar for the cap in the lathe, face, center drill, and drill the 3/16" hole 1/2" deep.

46. Turn the 45° angle 3/16" back.

47. Using the cutoff tool, cut the cap off at 3/8".

48. Turn cap end for end, CAREFULLY place the cap in the chuck, and drill the 1/4" hole 3/16" deep.

49. Place the 3/16"-diameter stock in the lathe, face to length, and put a radius on each end.

50. Slide the 3/16"-diameter handle into screw head and smash ends in press or vise so it will not fall out.

51. Place screw into "C" and put cap on end of screw; using a socket-head cap screw, insert the end of the screw into the 1/4" hole of cap and the other end against the side of "C" and tighten the screw. This will mushroom out the center-drilled hole in the end of the screw to keep it on the screw.

52. Evaluate and turn in.

Student Name: ______________________ Date Submitted: ____________

Class: ______________ Total Hours on Job: ____________

	Print Dimensions and Tolerances	Student's Inspection	Instructor's Inspection	Instructor Comments
1	Is the part free of burrs and does it meet finish requirements?			
2	Is the part properly identified?			
3	.750 ± .005 Thickness of C			
4	3.500 ± .005 Length of C			
5	2 ± .015 Height of C			
6	.625 ± .005 Threaded Side of C			
7	.625 ± .005 Bottom of C			
8	.625 ± .005 Angle Side of C			
9	.125 ± .005 Angle on C			You will NEED to use the comparator
10	.625 ± .005 Angle on C			You will NEED to use the comparator
11	.750 ± .005 Screw Head Length			You will NEED to use the comparator
12	Ø.625 ± .005 Head of Screw Diameter			
13	Ø.435 ± .002 Diameter of 7/16—14 Thread			
14	.12 ± .015 Diameter Behind Head, Unthreaded Portion of Screw			You will NEED to use the comparator
15	.375 ± .005 Length of cap			You will NEED to use the comparator
16	.187 ± .015 Flat on Cap			You will NEED to use the comparator
17	Ø.625 ± .005 Diameter of Cap			
18	45° ± 1/2° Angle on Cap			You will NEED to use the comparator
19	.125 ± .005 Radius Outside C			You will NEED to use the comparator

Project Grade: ____________

Lathe Dog

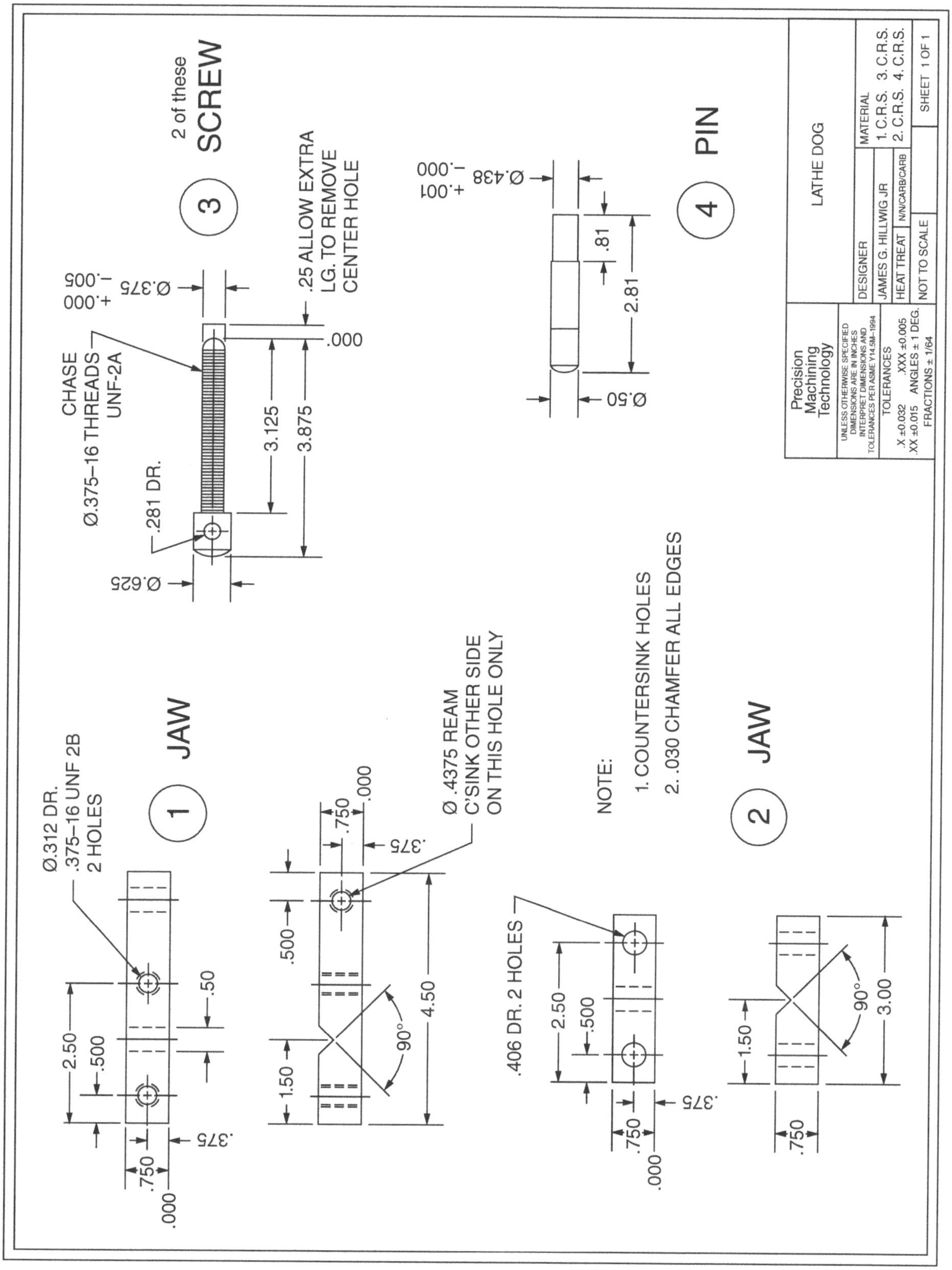
1 JAW
Ø.312 DR.
.375–16 UNF 2B
2 HOLES
Ø .4375 REAM
C'SINK OTHER SIDE
ON THIS HOLE ONLY
2 JAW
.406 DR. 2 HOLES
NOTE:
1. COUNTERSINK HOLES
2. .030 CHAMFER ALL EDGES
3 SCREW
2 of these
CHASE
Ø.375–16 THREADS
UNF-2A
.281 DR.
Ø.625
Ø.375 +.000 −.005
.25 ALLOW EXTRA
LG. TO REMOVE
CENTER HOLE
3.125
3.875
4 PIN
Ø.438 +.001 −.000
Ø.50
.81
2.81
Precision
Machining
Technology
UNLESS OTHERWISE SPECIFIED
DIMENSIONS ARE IN INCHES
INTERPRET DIMENSIONS AND
TOLERANCES PER ASME Y14.5M–1994
TOLERANCES
.X ±0.032 .XXX ±0.005
.XX ±0.015 ANGLES ± 1 DEG.
FRACTIONS ± 1/64
LATHE DOG
DESIGNER
JAMES G. HILLWIG JR
HEAT TREAT N/A/CARB/CARB
NOT TO SCALE
MATERIAL
1. C.R.S. 3. C.R.S.
2. C.R.S. 4. C.R.S.
SHEET 1 OF 1

NIMS Skills Practiced

- Job Process Planning
- Layout
- Turning Operations: Chucking
- Drilling and Tapping
- Surface Grinding and Safety
- Vertical Milling and Safety
- Squaring up a Block

Equipment

Power saw, milling machine, lathe, surface grinder

Tools Needed

Scribe
Dead blow hammer
Various drill bits
90° countersink
Edge finder
.4375 reamer
Tramming items for mill
Various types of turning, threading, and facing tools
6" steel rule
File
3/8-16 tap
Various sizes of endmills
Center drill
Grinding wheel
Adjustable mill angle
Layout dye
File card
Stamps
Parallels
Vise stop
Grinder vise
Thread triangles

Material Needed

- 3/4" square × 4-5/8" long CRS (Jaw Part #1)
- 3/4" square × 3-1/8" long CRS (Jaw Part #2)
- 2 @ 5/8" diameter × 4-1/8" long (Screws Part #3)
- 1/2" diameter × 3" long CRS (Pin Part #4)

Safety

1. Follow all shop safety rules given to you by your instructor and be sure to pass any safety tests given before using shop equipment.
2. Follow all safety rules for power equipment.
3. Be careful of sharp tools when using them.
4. Make sure the file has a handle on it.
5. Be sure all cutting tools are sharp and care is taken while handling them.
6. Be sure ways are locked on mill when necessary.
7. Be sure chuck is tightened and chuck wrench is never left in chuck.
8. Make sure stock is supported with center when necessary.
9. Ring grinding wheels before using them.
10. Make sure magnetic chuck is on when surface grinding.

Order of Operations

MILL

1. Look over blueprint to determine tolerances, tools, and material needed.
2. Gather tools needed, figure out RPMs for required tools, and write down their RPMs on sheet.
3. Measure, mark, and cut all stock needed.
4. Make sure milling machine is in tram.
5. Deburr all over.
6. Place piece in mill vise on parallels with sawed end sticking out; tap down with dead blow hammer to seat part on parallels, then mill just enough off the end to clean it up.
7. Deburr the end you just milled, flip part 180° on parallels in vise, tap down with dead blow hammer to seat part on parallels, and mill to length.
8. Repeat with part #2.
9. Lay out parts.
10. Place part #1 in vise on parallels against vise stop, tap down with dead blow hammer to seat part on parallels, and set vise stop on left side of part.
11. Insert edge finder in collet, set RPMs to 1100, and edge find the X-axis and the Y-axis. Be sure to allow for 1/2 grind stock on the X-axis.
    a. X-axis should read −.1075 (this includes 1/2 of the edge finder and 1/2 of the grind stock).
    b. Y-axis should read .1000 (with the tolerance on the print, the 3/4" stock size should leave plenty of room for grinding).
12. Place center drill in chuck and move to hole locations.
13. Lower center drill to make sure it aligns with layout lines.
14. If location is correct, center drill hole at 1200 RPMs, change to drill, set the RPMs, and drill through part.
15. Repeat with each hole on parts #1 and #2.
16. Countersink holes where indicated on print.
17. You can power tap the two tapped holes in part #1 or use a tap stand.
18. Lay out the lines for the angle and set parts together on an adjustable mill angle and mill the angle.

SURFACE GRINDER

MAKE SURE MAGNETIC CHUCK IS ON WHEN SURFACE GRINDING!

19. Surface grind all sides square and parallel. Deburr after each side is ground.
20. Surface grind the face to clean it up, deburr, then flip the part to grind opposite side.
21. Deburr all edges and place in grinder vise on parallels with ground sides against jaws, leaving one edge sticking out of side of vise so you can turn vise on its side.
22. Grind top of part to clean up and turn vise on side without removing or disturbing part while in vise and grind the end just enough to clean it up.
23. Remove from vise, deburr all edges, place ground side down on chuck with square blocking on each side, and grind bottom to clean it up.
24. Deburr and place last side up with square blocking on each side and clean up surface.
25. Now check sizes and hole location; grind them into location and part to size. Be sure to deburr and block part when needed.
26. Repeat with part #2.
27. Lay flat on grinder chuck and grind a .030-by-45° chamfer on all edges.

LATHE

28. Chuck 5/8" diameter stock for screw in lathe face until end is clean.
29. Mark faced end at .750; this will be the head of the screw.
30. File the convex on the face of the screw.

31. Turn part end for end, face to length, and center drill using a #3 center drill. The center-drilled hole remains in the part for turning and threading, and then it will be removed.
32. Turn .375 diameter back 3-3/8" to the .750 mark of the head.
33. Chase the thread on the screw.
34. Face off center drill and take to length, and then file the radius on the end.
35. Set screw in mill or drill press and drill the .281 hole in the head.
36. Place the 1/2"-diameter stock in the lathe, face both ends, and take to length.
37. Put a radius on end.
38. Flip end for end and turn .438 diameter back .81.
39. Place pin in part #1 through the non-countersunk side.
40. Place the pin radius on a piece of aluminum on top of an anvil, and hammer the pin into the countersink.
41. Place in grinder vise with the mushroomed end sticking up and surface grind until pin and part #1 are blended. (If done correctly, they should look like one piece.)
42. Assemble the rest of the part.
43. Evaluate and turn in.

Student Name: ______________________ Date Submitted: ____________

Class: ______________ Total Hours on Job: ____________

	Print Dimensions and Tolerances	Student's Inspection	Instructor's Inspection	Instructor Comments
1	Is the part free of burrs and does it meet finish requirements?			
2	Is the part properly identified?			
3	**{Part #1 – JAW}** 4.50 ± .015			
4	.750 ± .005 Square			
5	Hole Location .500 ± .005			
6	Hole Location 2.50 ± .015			
7	(.312 Drill/Tap) Ø.375—16 Threads			
8	90° ± 1° Angle Width .50 ± .015			
9	90° ± 1° Angle Location 1.50 ± .015			
10	**{Part #2 – JAW}** 3.00 ± .015			
11	.750 ± .005 Square			
12	Hole Location .500 ± .005			
13	Hole Location 2.50 ± .015			
14	Holes .406 ± .005			
15	90° ± 1° Angle Width .50 ± .015			
16	90° ± 1° Angle Location 1.50 ± .015			
17	Are the holes countersunk?			
18	**{Part #3 – Screw 1}** 3.875 ± .005			
19	3.125 ± .005			
20	Ø.375 ± .000 − .005			
21	Ø.375 – 16 Threads			
22	Ø.625 Diameter .750 ± .005			
23	Ø.281 Hole ± .005			

(Continued)

	Print Dimensions and Tolerances	Student's Inspection	Instructor's Inspection	Instructor Comments
24	**(Part #3 – Screw 2)** 3.875 ± .005			
25	3.125 ± .005			
26	3.125 ± .005			
27	Ø.375 ± .000 − .005			
28	Ø.375 – 16 Threads			
29	Ø.625 Diameter .750 ± .005			
30	Ø.281 Hole ± .005			
31	**(Part #4 – Pin)** Ø.05 ± .015			
32	Does the radius look like the print?			
33	Is the pin properly ground to blend into the part?			
34	Are there 45°-by-.030 chamfers on the edges of the parts?			

Project Grade: ____________________

Micrometer Stand

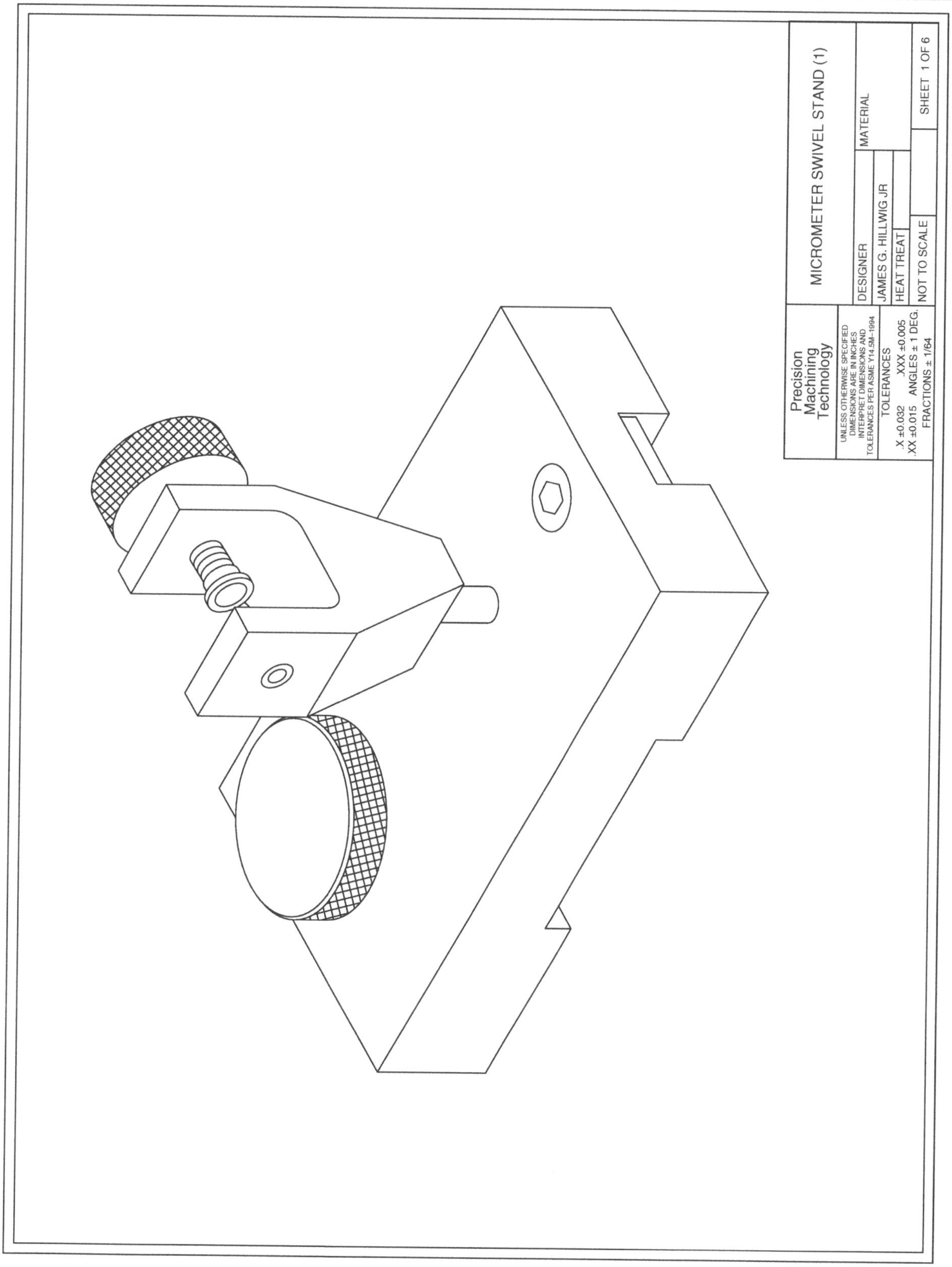

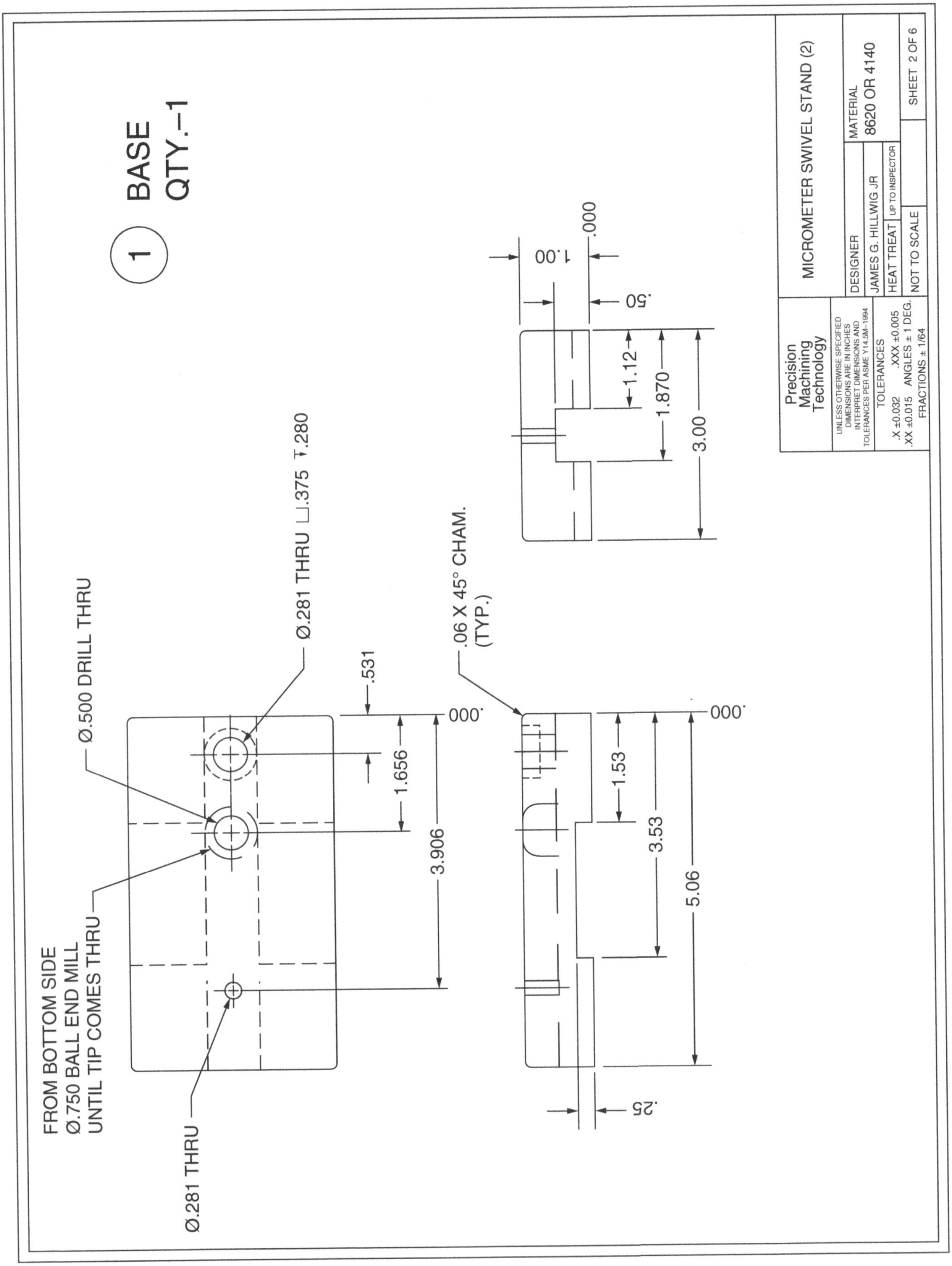

1 BASE
QTY.–1
Ø.500 DRILL THRU
Ø.281 THRU ⌴.375 ↧.280
.06 X 45° CHAM.
(TYP.)
FROM BOTTOM SIDE
Ø.750 BALL END MILL
UNTIL TIP COMES THRU
Ø.281 THRU
.000
.531
1.656
3.906
.000
1.53
3.53
5.06
.25
1.00
.000
.50
1.12
1.870
3.00
Precision
Machining
Technology
UNLESS OTHERWISE SPECIFIED
DIMENSIONS ARE IN INCHES
INTERPRET DIMENSIONS AND
TOLERANCES PER ASME Y14.5M-1994
TOLERANCES
.X ±0.032
.XX ±0.015
.XXX ±0.005
ANGLES ± 1 DEG.
FRACTIONS ± 1/64
MICROMETER SWIVEL STAND (2)
DESIGNER
JAMES G. HILLWIG JR
MATERIAL
8620 OR 4140
HEAT TREAT
UP TO INSPECTOR
NOT TO SCALE
SHEET 2 OF 6

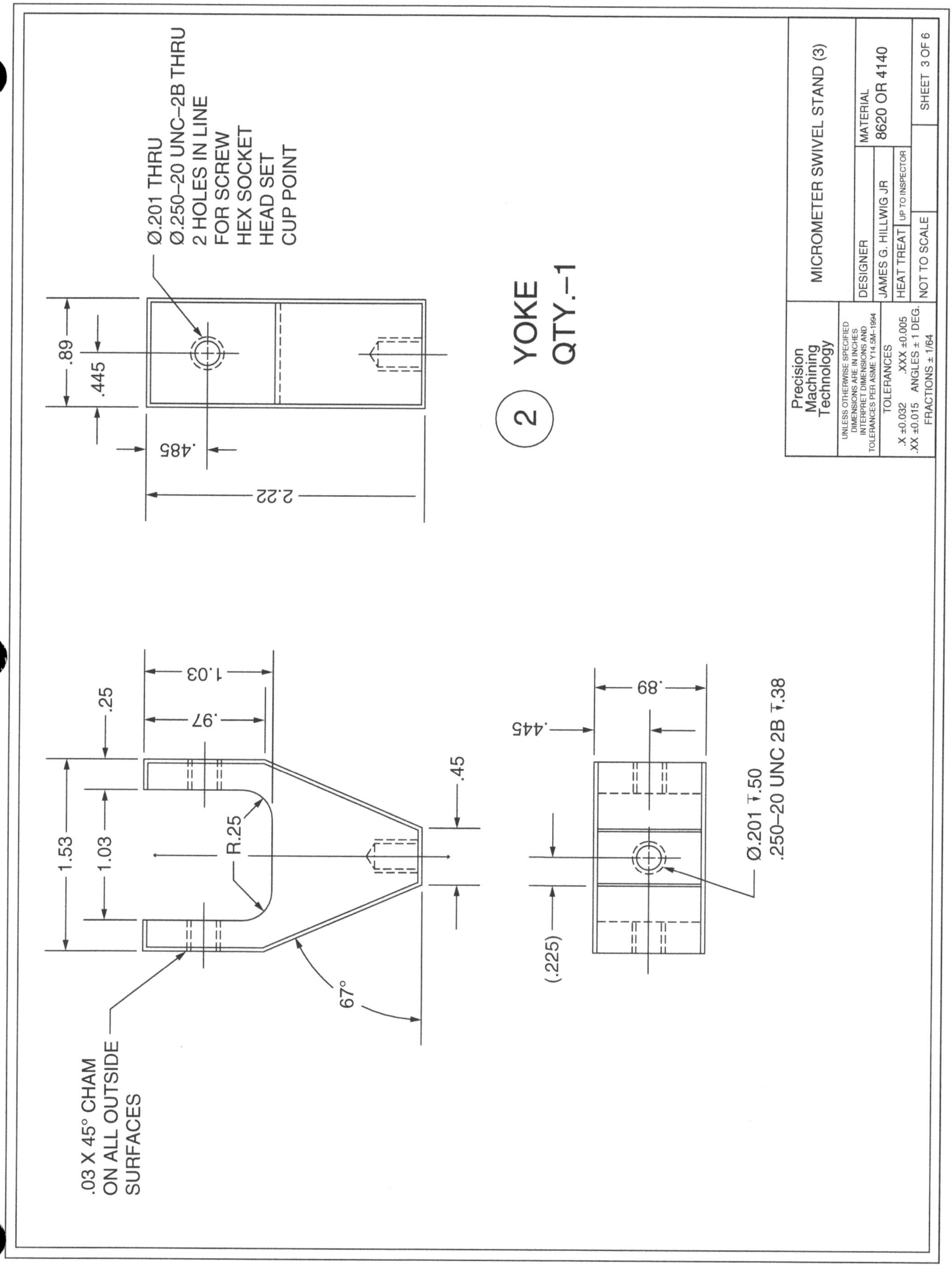

Ø.201 THRU
Ø.250–20 UNC–2B THRU
2 HOLES IN LINE
FOR SCREW
HEX SOCKET
HEAD SET
CUP POINT
.89
.445
.485
2.22
2 YOKE
QTY.–1
MICROMETER SWIVEL STAND (3)
Precision Machining Technology
UNLESS OTHERWISE SPECIFIED DIMENSIONS ARE IN INCHES INTERPRET DIMENSIONS AND TOLERANCES PER ASME Y14.5M–1994
TOLERANCES
.X ±0.032 .XXX ±0.005
.XX ±0.015 ANGLES ± 1 DEG.
FRACTIONS ± 1/64
DESIGNER
JAMES G. HILLWIG JR
MATERIAL
8620 OR 4140
HEAT TREAT
UP TO INSPECTOR
NOT TO SCALE
SHEET 3 OF 6
.25
1.03
.97
1.53
1.03
R.25
.45
67°
.03 X 45° CHAM
ON ALL OUTSIDE
SURFACES
.89
.445
(.225)
Ø.201 ↧.50
.250–20 UNC 2B ↧.38

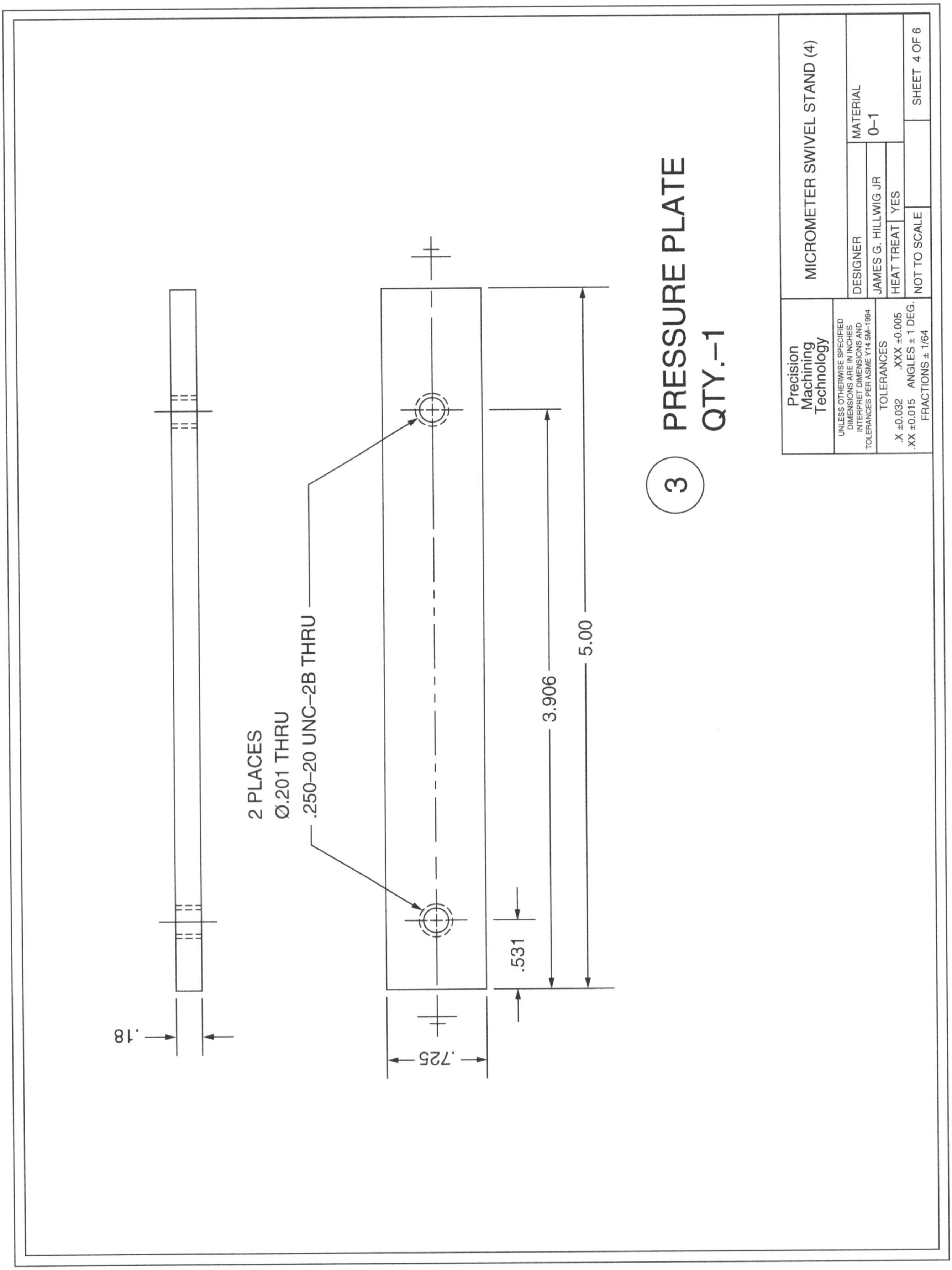
MICROMETER SWIVEL STAND (4)
Precision Machining Technology
UNLESS OTHERWISE SPECIFIED DIMENSIONS ARE IN INCHES INTERPRET DIMENSIONS AND TOLERANCES PER ASME Y14.5M-1994
TOLERANCES
.X ±0.032 .XXX ±0.005
.XX ±0.015 ANGLES ± 1 DEG.
FRACTIONS ± 1/64
DESIGNER
JAMES G. HILLWIG JR
MATERIAL
0–1
HEAT TREAT YES
NOT TO SCALE
SHEET 4 OF 6
3
PRESSURE PLATE
QTY.–1
2 PLACES
Ø.201 THRU
.250–20 UNC–2B THRU
5.00
3.906
.531
.725
.18

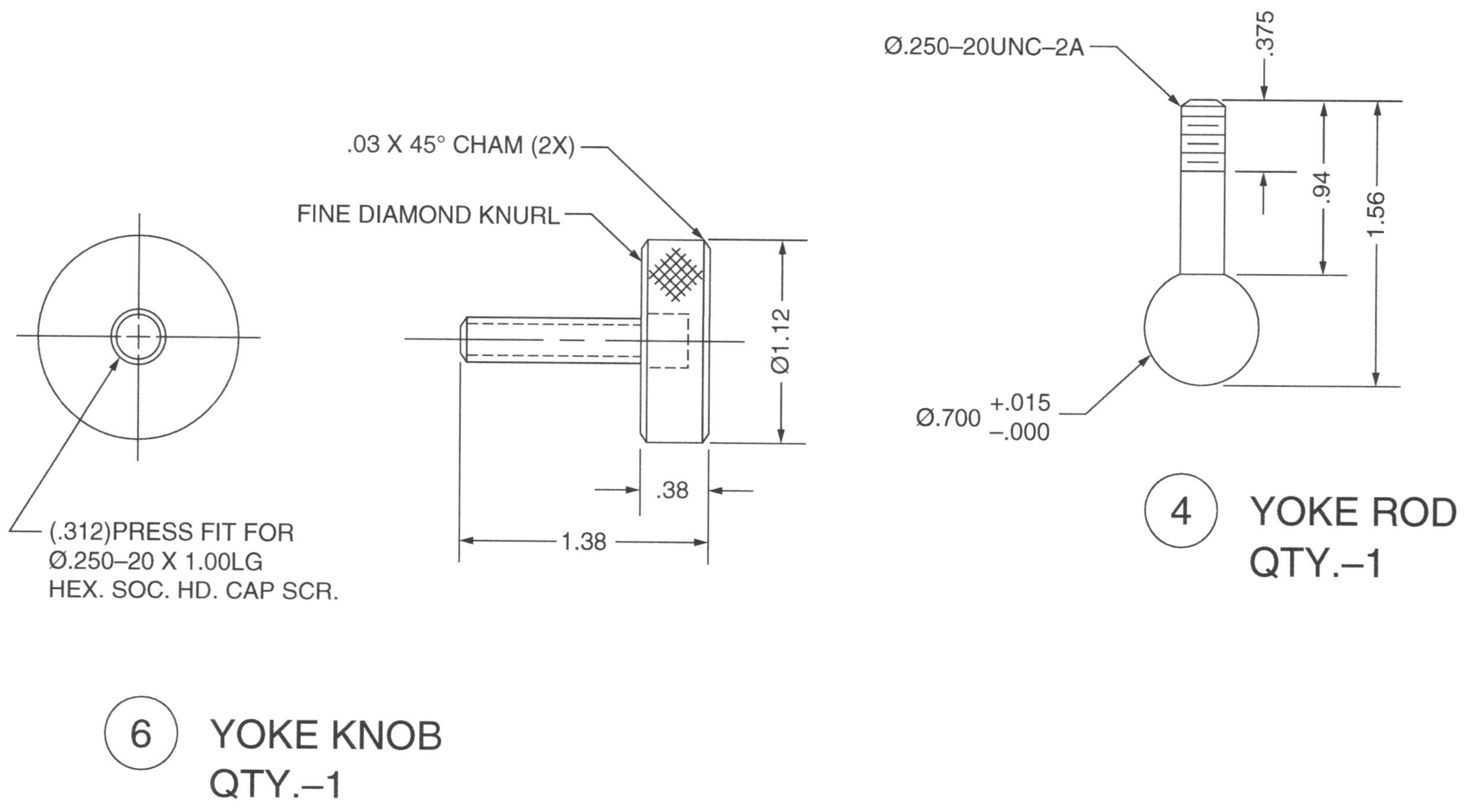

Precision Machining Technology	MICROMETER SWIVEL STAND (5)	
UNLESS OTHERWISE SPECIFIED DIMENSIONS ARE IN INCHES INTERPRET DIMENSIONS AND TOLERANCES PER ASME Y14.5M–1994	DESIGNER JAMES G. HILLWIG JR	MATERIAL BRASS 0–1
TOLERANCES .X ±0.032 .XXX ±0.005 .XX ±0.015 ANGLES ± 1 DEG. FRACTIONS ± 1/64	HEAT TREAT YES	
	NOT TO SCALE	SHEET 5 OF 6

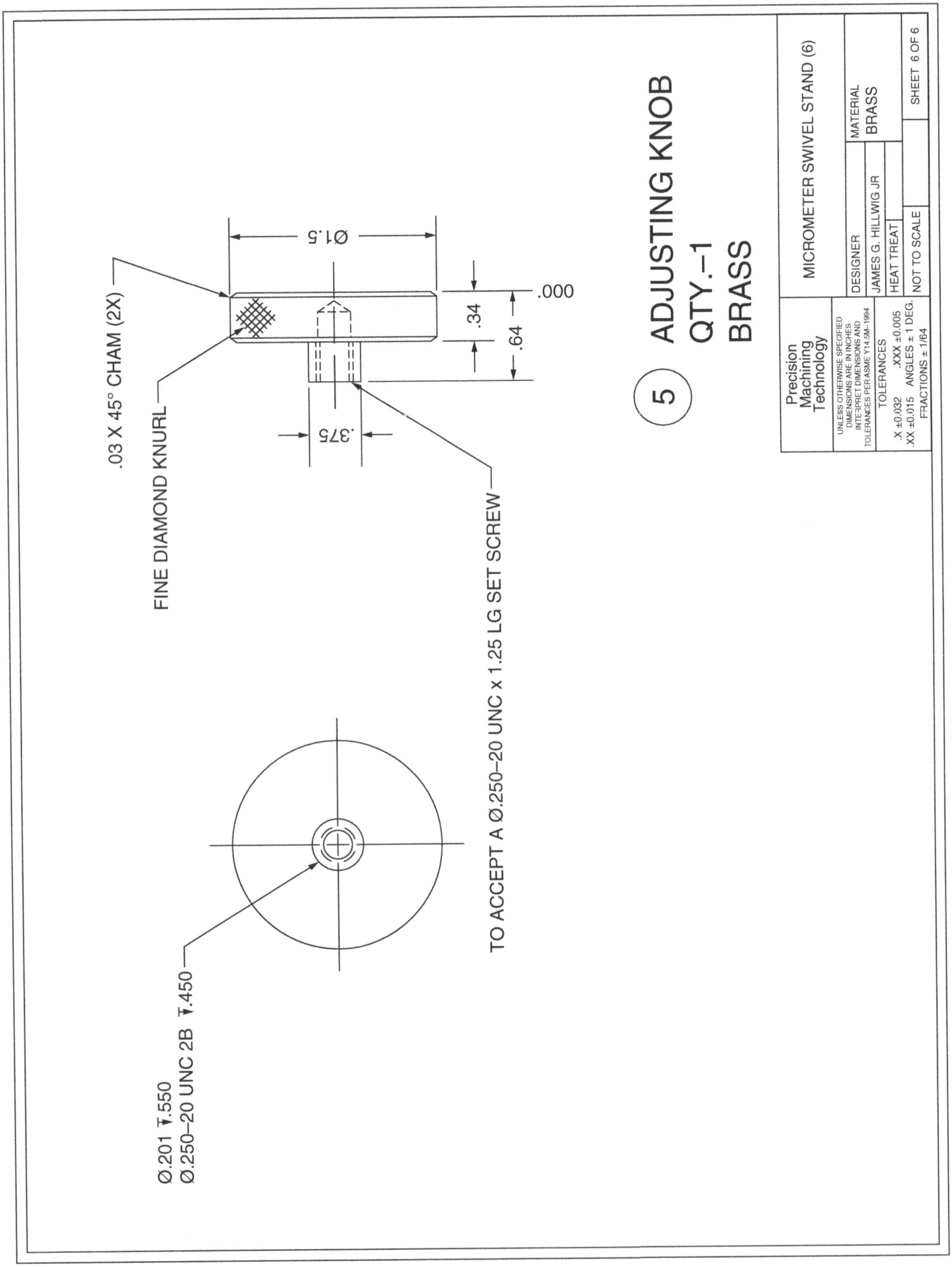

5 ADJUSTING KNOB
QTY.–1
BRASS
.03 X 45° CHAM (2X)
FINE DIAMOND KNURL
Ø1.5
.000
.34
.64
.375
TO ACCEPT A Ø.250–20 UNC x 1.25 LG SET SCREW
Ø.201 ↧.550
Ø.250–20 UNC 2B ↧.450
Precision Machining Technology
UNLESS OTHERWISE SPECIFIED DIMENSIONS ARE IN INCHES INTERPRET DIMENSIONS AND TOLERANCES PER ASME Y14.5M–1994
TOLERANCES
.X ±0.032 .XXX ±0.005
.XX ±0.015 ANGLES ± 1 DEG.
FRACTIONS ± 1/64
MICROMETER SWIVEL STAND (6)
DESIGNER
JAMES G. HILLWIG JR
MATERIAL
BRASS
HEAT TREAT
NOT TO SCALE
SHEET 6 OF 6

NIMS Skills Practiced

- Job Process Planning
- Layout
- Turning Operations: Chucking
- Drilling and Tapping
- Surface Grinding and Safety
- Vertical Milling and Safety
- Squaring Up a Block
- CNC Milling
- CNC Turning

Equipment

Power saw, milling machine, lathe, CNC mill, CNC lathe, surface grinder

Tools Needed

Scribe	6" steel rule	Layout dye
Dead blow hammer	File	File card
Various drill bits	1/4-20 tap	Stamps
90° countersink	Various sizes of endmills	Parallels
Edge finder	Center drill	Vise stop
Grinding wheel	Grinder vise	Tramming items for mill
Adjustable mill angle	Thread triangles	Various types of turning, threading, and facing tools

Material Needed

- 1" × 3" × 5.185" long CRS (Base Part #1)
- 1-1/8" × 1-3/4" × 2-3/8" long CRS (Yoke Part #2)
- .1875 × .750 × 5-1/8" long 0-1 (Pressure Plate Part #3)
- 3/4" diameter × 1-3/4" long 0-1 (Yoke Knob Part #4)
- 1.125 diameter × .38 brass (Yoke Knob Part #6—use enough stock to safely hold part)
- 1.5 diameter × .64 brass (Adjusting Knob Part #5—use enough stock to safely hold part)

Safety

1. Follow all shop safety rules given to you by your instructor and be sure to pass any safety tests given before using shop equipment.
2. Follow all safety rules for power equipment.
3. Be careful of sharp tools when using them.
4. Make sure the file has a handle on it.
5. Be sure all cutting tools are sharp and care is taken while handling them.
6. Be sure ways are locked on mill when necessary.
7. Be sure chuck is tightened and chuck wrench is never left in chuck.
8. Make sure stock is supported with center when necessary.
9. Ring grinding wheels before using them.
10. Make sure magnetic chuck is on when surface grinding.
11. Follow all CNC rules and safety guidelines for operation and programming.

Order of Operations

BASE

MILL

1. Look over blueprint to determine tolerances, tools, and materials needed.
2. Gather tools needed, figure out RPMs for required tools, and write down their RPMs on sheet.
3. Measure, mark, and cut all stock needed.
4. Make sure milling machine is in tram.
5. Deburr all over.
6. Place piece in mill vise on parallels, square it up, and take to size.
7. Deburr part.
8. Lay out part.
9. Edge find part.
10. Center drill and drill holes.
11. From top of part, counterbore hole where indicated.
12. On bottom of part, using a 3/4" ball endmill, drill in .500 hole until tip of endmill breaks through .500 hole. This will create the radius for the ball to rotate on.
13. With the part still in vise, mill out the two slots.
14. Countersink holes where indicated on print.
15. Send for heat treatment.

YOKE

CNC MILL

1. Following your instructor's directions, cut stock to size and write the program.
2. Deburr the part and place in the CNC mill.
3. Load program and tools and follow your instructor's directions to cut out part.
4. You can drill and tap 1/4-20 holes using a manual mill or program the CNC.
5. Send for heat treatment.

PRESSURE PLATE

MILL

1. Cut, deburr, square, and mill part to size.
2. Drill and tap both holes for 1/4-20 screws.
3. Send for heat treatment.

YOKE ROD

CNC LATHE

1. Following your instructor's directions, cut stock to size recommended by your instructor and write the program.
2. Deburr the part and place in the CNC lathe.
3. Load program and tools and follow your instructor's directions to cut out part.
4. Send for heat treatment.

YOKE KNOB

CNC OR MANUAL LATHE

1. If using the CNC lathe, follow steps above.
2. If using manual lathe, cut a bar of brass approximately 12" long (so you can safely hold in chuck), face one end, center drill, and drill to print dimensions.
3. Slide a few inches out of chuck and knurl.
4. Cut off part. Partway through cutoff, stop. Using a file, file chamfers on both sides of knob. Continue cutting off part.
5. Press screw into knob and polish top.

ADJUSTING KNOB

CNC OR MANUAL LATHE

1. If using the CNC lathe, follow steps above.
2. If using manual lathe, cut a bar of brass approximately 12" long (so you can safely hold in chuck), face one end, center drill, and drill to print dimensions.
3. Slide a few inches out of chuck and knurl.
4. Turn down .375 shoulder.
5. Following your instructor's directions, you can tap the hole on the lathe.
6. Cut off part. Partway through cutoff, stop. Using a file, file chamfers on both sides of knob. Continue cutting off part.
7. Loctite screw into knob and polish top.

SURFACE GRINDER

MAKE SURE MAGNETIC CHUCK IS ON WHEN SURFACE GRINDING!

1. Surface grind all sides square and parallel on each part. Deburr after each side is ground.
2. Lay flat on grinder chuck and grind chamfer on all edges where indicated on print.

The part looks really nice if it is sent out for blackening when all of the grinding is complete. On the two 1/4-20 screws that go into the top of the yoke, if you get a couple of black rubber-flanged sink washers, run a tap through the hole in the washers and thread them on the inside of the screws. They will clamp your micrometers without the screws marring the surface of the micrometers.

Student Name: ______________________ Date Submitted: ____________

Class: ____________ Total Hours on Job: ____________

	Print Dimensions and Tolerances	Student's Inspection	Instructor's Inspection	Instructor Comments
1	Is the part free of burrs and does it meet finish requirements?			
2	Is the part properly identified?			
3	**{Base Length}** 5.06 ± .015			
4	**{Base Width}** 3.00 ± .015			
5	**{Base Height}** 1.00 ± .015			
6	**{Hole Location}** 3.906 ± .005			
7	**{Hole Location}** 1.656 ± .005			
8	**{Hole Location}** .531 ± .005			
9	**{Base – Slot Width}** 2.00 ± .015 × .25 ± .015			
10	**{Base – Slot Width}** .75 ± .015 × .50 ± .015			
11	**{Yoke Length}** 2.22 ± .015			
12	**{Yoke Width}** 1.53 ± .015			
13	**{Yoke Height}** .89 ± .015			
14	**{Yoke Opening}** 1.03 ± .015			
15	**{Yoke Arms}** .25 ± .015			
16	**{Yoke Angle}** 67° ± 1/2°			
17	**{Pressure Plate Length}** 5.00 ± .015			
18	**{Pressure Plate Width}** .725 ± .005			
19	**{Pressure Plate Height}** .18 ± .015			
20	**{Hole Location}** 3.906 ± .005			

Project Grade: ____________

Tapping Block

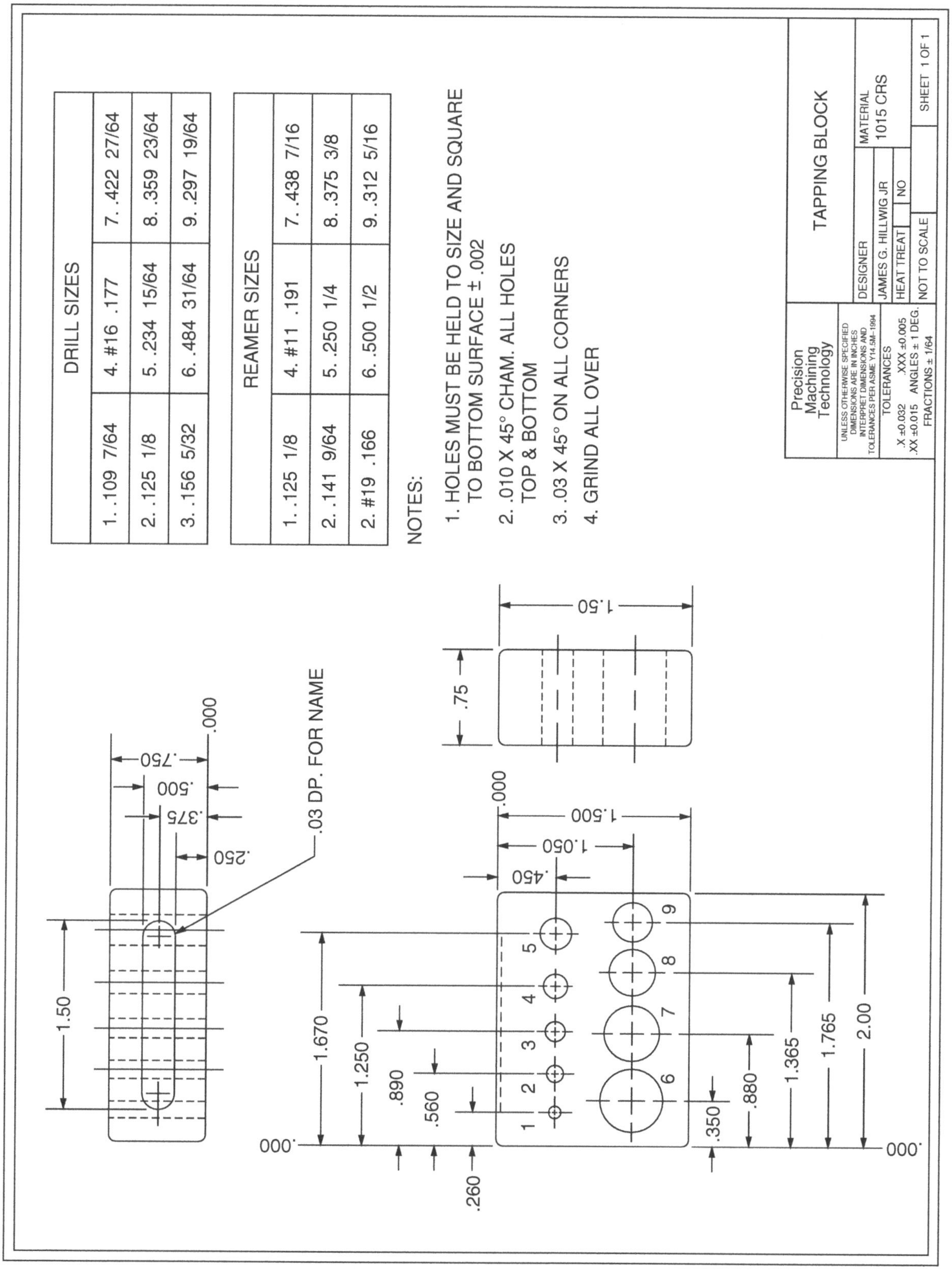

NIMS Skills Practiced

- Job Process Planning
- Layout
- Surface Grinding and Safety
- Vertical Milling and Safety
- Squaring up a Block

Equipment

Power saw, milling machine, surface grinder

Tools Needed

Scribe
Hammer
Various sizes of drill bits
90° countersink
Edge finder
Machinist square
Dead blow hammer
6″ steel rule
File
Various sizes of reamers
Various sizes of endmills
Center drills
Grinding wheel
Tramming items for mill
Layout dye
File card
Stamps
Parallels
Vise stop
Grinder vise

Material Needed

- 2.1 × 1.5 × .75 mild steel

Safety

1. Follow all shop safety rules given to you by your instructor and be sure to pass any safety tests given before using shop equipment.
2. Follow all safety rules for power equipment.
3. Be careful of sharp tools when using them.
4. Make sure the file has a handle on it.
5. Be sure all cutting tools are sharp and care is taken while handling them.
6. Be sure ways are locked on mill when necessary.
7. Ring grinding wheels before using them.
8. Make sure magnetic chuck is on when surface grinding.

Order of Operations

1. Look over blueprint to determine tolerances, tools, and material needed.
2. Gather tools needed, figure out RPMs for required tools, and write down their RPMs on sheet.
3. Measure, mark, and cut stock 3/4″ thick × 1 1/2″ wide; saw 2-1/8″ long.
4. Make sure milling machine is in tram.
5. Deburr all over; check squareness on rolled sides from steel mill.
6. Place piece in mill vise with sawed end up, placing machinist square next to part, and mill just enough off the end to clean it up.
7. Deburr the end you just milled, place the milled end on the bottom of the vise, turn or rotate piece 90° keeping the milled end on the bottom of the vise, and use the machinist square again to mill just enough off to clean this end up.

8. Deburr the end you just milled and place in the bottom of vise and take a light cut off the part; this will be the first end you cut.
9. If you did this correctly, the ends will be square—be sure to check them.
10. Mill part 2.015 long, deburr, and re-check for squareness.
11. Paint part with layout dye, then use height gage to lay out hole locations, being sure to leave grind stock on each end.
12. Place part in vise on parallels against vise stop; tap down with dead blow hammer to seat part on parallels.
13. Insert edge finder in collet, set RPMs to 1,100, and edge find the X-axis and the Y-axis. Be sure to allow for 1/2 grind stock on the X-axis.
    a. X-axis should read −.1075 (this includes 1/2 of the grind stock).
    b. Y-axis should read .1000.
14. Place center drill in chuck and move to −.450 on Y-axis and .260 on X-axis.
15. Lower center drill to make sure it aligns with layout lines.
16. If location is correct, center drill hole at 1,200 RPMs, change to 7/64 drill, set the RPMs, and drill through part; then insert 1/8 reamer, set RPMs, and ream hole through.
17. Move X-axis to .560, put center drill back in, lower to part, and check location to layout lines; if correct, set RPMs and center drill hole. Change to 7/64 drill, set the RPMs, and drill through part—this is called step drilling; then insert 1/8 drill, set RPMs, and drill through part. Insert 9/64 reamer, set RPMs, and ream through hole.
18. Repeat this process for each hole—step drilling each hole, drilling hole size, then reaming hole.
19. Remove part from vise and deburr holes.
20. Return part to vise against vise stop and countersink each hole, both sides. Remember to set the RPMs for the countersink.
21. Lay out name pocket, 3/8 wide × 1″ long × .030 deep, centered on part.
22. Using a 3/8 endmill, set RPMs, locate pocket, and mill to print dimensions; remove part, deburr, place on anvil, and, using steel stamps, stamp your name in pocket.

MAKE SURE MAGNETIC CHUCK IS ON WHEN SURFACE GRINDING!

23. Surface grind all sides square and parallel. Deburr after each side is ground.
24. Surface grind the face to clean it up, deburr, then flip the part to grind bottom.
25. Deburr all edges, and place all the way down in grinder vise with ground sides against jaws, leaving edge with name pocket sticking out of top of vise. Leave one edge sticking out of side of vise, so you can turn vise on its side.
26. Grind top of part to clean up and turn vise on side without removing or disturbing part while in vise just to clean it up.
27. Remove from vise, deburr all edges, and place name pocket side down on chuck with square blocking on each side and grind side to clean it up.
28. Deburr and place last side up with square blocking on each side and clean up surface.
29. Now check hole locations and grind them into location and part to size. Be sure to deburr and block part when needed.
30. Grind a .030 × 45° chamfer on all edges.
31. Inspect the part to print dimensions; record on inspection sheet and turn in.

How to Determine the RPMs

Tool Being Used	SFPM	Formula	RPM
Edge Finder	80	Given	1,000
#2 Center Drill	80	Given	1,200
#3 Center Drill	80	Given	1,100
Drill 7/6	80	$\frac{4 \times 80}{.109}$	
Drill 1/8	80	$\frac{4 \times 80}{.125}$	
Drill 5/32	80	$\frac{4 \times 80}{.156}$	
Drill #16	80	$\frac{4 \times 80}{.177}$	
Drill 15/64	80	$\frac{4 \times 80}{.234}$	
Drill 31/64	80	$\frac{4 \times 80}{.484}$	
Drill 27/64	80	$\frac{4 \times 80}{.422}$	
Drill 23/64	80	$\frac{4 \times 80}{.359}$	
Drill 19/64	80	$\frac{4 \times 80}{.297}$	

How to Determine the RPMs for Reamers

Tool Being Used	SFPM	Formula	RPM
Reamer 1/8	80	$\frac{4 \times 80}{.125}$	Divide RPM in half for reamers
Reamer 9/64	80	$\frac{4 \times 80}{.141}$	Divide RPM in half for reamers
Reamer #19	80	$\frac{4 \times 80}{.166}$	Divide RPM in half for reamers
Reamer #11	80	$\frac{4 \times 80}{.191}$	Divide RPM in half for reamers
Reamer 1/4	80	$\frac{4 \times 80}{.250}$	Divide RPM in half for reamers
Reamer 1/2	80	$\frac{4 \times 80}{.500}$	Divide RPM in half for reamers
Reamer 7/16	80	$\frac{4 \times 80}{.438}$	Divide RPM in half for reamers
Reamer 3/8	80	$\frac{4 \times 80}{.375}$	Divide RPM in half for reamers
Reamer 5/16	80	$\frac{4 \times 80}{.312}$	Divide RPM in half for reamers
90° Countersink	80	$\frac{4 \times 80}{\text{Tool } \varnothing}$	

Student Name: ______________________ Date Submitted: __________

Class: __________ Total Hours on Job: __________

	Print Dimensions and Tolerances	Student's Inspection	Instructor's Inspection	Instructor Comments
1	Is the part free of burrs and does it meet finish requirements?			
2	Is the part properly identified?			
3	Front View 2.00 ± .015			
4	Side View 1.50 ± .015			
5	Side View .75 ± .015			
6	Front View, Hole #1 1/8 Reamer ± .015			
7	Front View Hole #1 .260 ± .005 by .450 ± .005			
	Top Row of Holes			
8	Front View Holes #2 through #5 Reamed to Dimensions ± .002			
9	Front View Hole #2 .560 ± .005 by .450 ± .005			
10	Front View Hole #3 .890 ± .005 by .450 ± .005			
11	Front View Hole #4 1.250 ± .005 by .450 ± .005			
12	Front View Hole #5 1.670 ± .005 by .450 ± .005			
	Bottom Row of Holes			
13	Front View Holes #6 through #9 Reamed to Dimensions ± .002			
14	Front View Hole #6 .350 ± .005 by 1.050 ± .005			
15	Front View Hole #7 .880 ± .005 by 1.050 ± .005			
16	Front View Hole #8 1.365 ± .005 by 1.050 ± .005			
17	Front View Hole #9 1.765 ± .005 by 1.050 ± .005			
18	Grind Finish			
19	Chamfers on Outside of Part and Holes			

Project Grade: __________

Bench Block

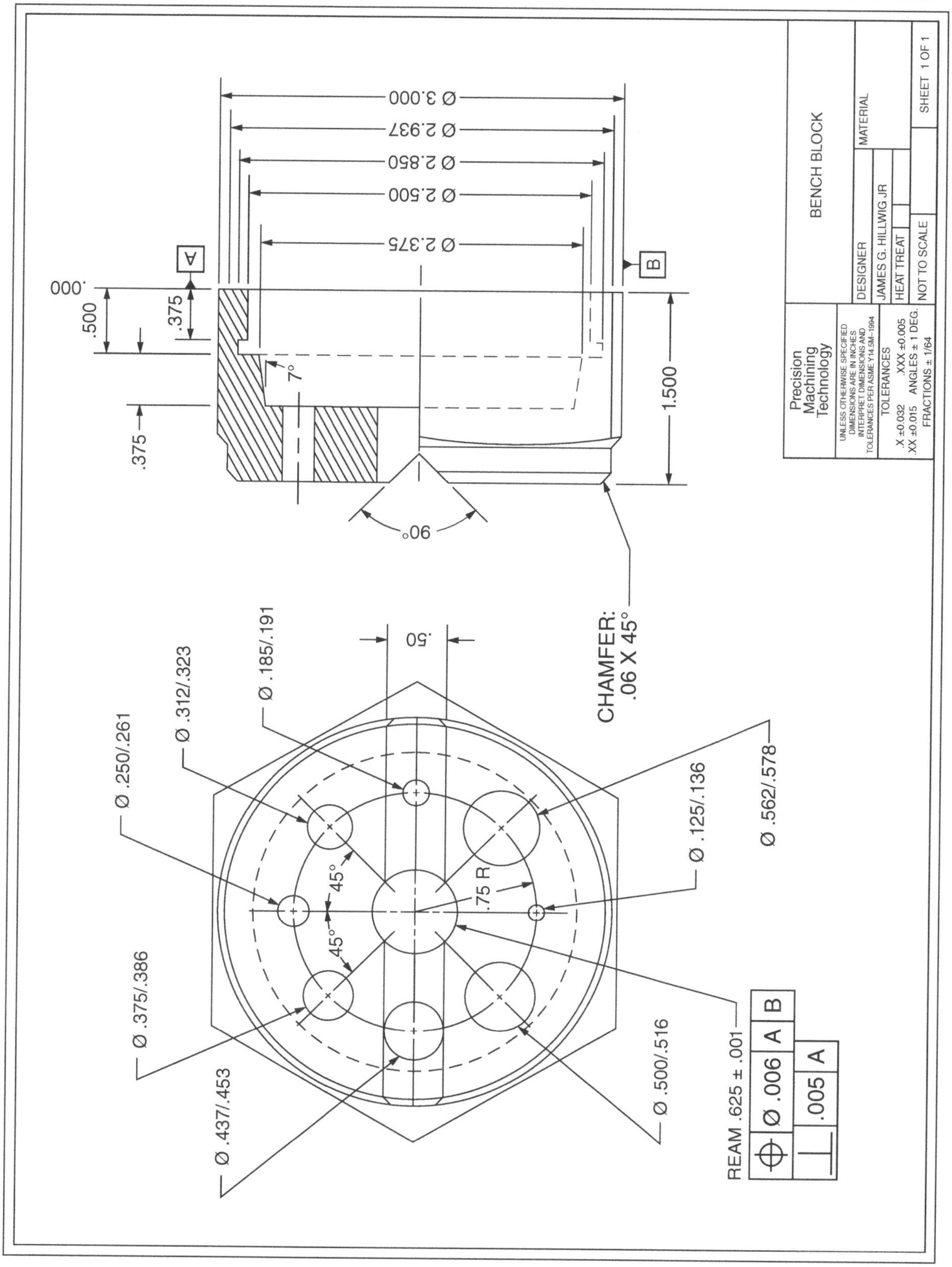
Ø 3.000
Ø 2.937
Ø 2.850
Ø 2.500
Ø 2.375
A
B
.000
.500
.375
.375
7°
1.500
90°
CHAMFER:
.06 X 45°
.50
Ø .185/.191
Ø .312/.323
Ø .250/.261
Ø .375/.386
Ø .437/.453
Ø .500/.516
Ø .125/.136
Ø .562/.578
.75 R
45°
45°
REAM .625 ± .001
Ø .006 A B
.005 A
Precision Machining Technology
UNLESS OTHERWISE SPECIFIED DIMENSIONS ARE IN INCHES INTERPRET DIMENSIONS AND TOLERANCES PER ASME Y14.5M-1994
TOLERANCES
.X ±0.032 .XXX ±0.005
.XX ±0.015 ANGLES ± 1 DEG.
FRACTIONS ± 1/64
BENCH BLOCK
DESIGNER
JAMES G. HILLWIG JR
MATERIAL
HEAT TREAT
NOT TO SCALE
SHEET 1 OF 1

NIMS Skills Practiced

- Job Process Planning
- Turning Operations: Chucking
- Milling
- Drilling
- Grinding

Equipment

Power saw, lathe, bench grinder, mill and surface grinders

Tools Needed

Marker
Various drills
Hammer
Various types of turning, facing, and boring tools
6″ steel rule
Adjustable mill angle
Layout dye
5/8″ reamer
Center drills
Center punch
Dividers

Material Needed

- Hex stock (CRS): 3″ hex × 1-5/8″ long

Safety

1. Follow all shop safety rules given to you by your instructor and be sure to pass any safety tests given before using shop equipment.
2. Follow all safety rules for power equipment.
3. Be careful of sharp tools when using them.
4. Make sure the files have a handle on them.
5. Be sure all cutting tools are sharp and care is taken while handling them.
6. Be sure ways are locked on mill when necessary.
7. Be sure chuck is tightened and chuck wrench is never left in chuck.
8. Make sure stock is supported with center when necessary.
9. Ring grinding wheels before using them.
10. Make sure magnetic chuck is on when surface grinding.

Order of Operations

1. Look over blueprint to determine tolerances, tools, and material needed.
2. Gather tools needed, figure out RPMs for required stock diameters and drills used, and write down their RPMs on sheet.
3. Measure and cut stock sizes needed.
4. Chuck in lathe, face, and center drill top.
5. Step drill the hole out to 19/32, then ream through with a 5/8″ reamer.
6. Turn 2-15/16″ diameter back 3/8″.
7. Turn a 45° chamfer × 1/16″ on front.
8. Turn end for end in lathe and step drill as close to 2″ as you can.

9. Remove the rest with a boring tool to 2″ inside diameter 7/8″ deep.
10. Bore 2-1/2″ diameter 1/2″ in from end.
11. Now with the boring tool, turn the inside angle of 7° in the bottom 3/8″ of hole until the large diameter is 2-3/8″.
12. Lay out to holes on the top to be drilled, center punch the holes, and drill them on the drill press—first with a center drill then the step drill or drill if needed and then to size.
13. Lay out the 45° angle marks on the top, place part in mill vise on adjustable mill angle, and mill the angle.
14. Mill name pocket in center of one of the flats and stamp name in it.
15. Deburr part and place on surface grinder and grind top just to clean it up.
16. Evaluate and turn in.

Student Name: ______________________ Date Submitted: ____________

Class: ______________ Total Hours on Job: ____________

	Print Dimensions and Tolerances	Student's Inspection	Instructor's Inspection	Instructor Comments
1	Is the part free of burrs and does it meet finish requirements?			
2	Is the part properly identified?			
3	**{Height}** 1-1/2" ± .015			
4	**{Hole}** 1/4" ± .015			
5	**{Hole}** 5/16" ± .015			
6	**{Hole}** 3/16" ± .015			
7	**{Hole}** 9/16" ± .015			
8	**{Hole}** 1/8" ± .015			
9	**{Hole}** 1/2" ± .015			
10	**{Hole}** 7/16" ± .015			
11	**{Hole}** 3/8" ± .015			
12	**{Hole}** 5/8" ± .015			
13	**{90° Angle}** 7/16" Wide ± .015			
14	Holes on a 7/8" Radius			
15	Holes 45° Apart			
16	**{Top Step}** 2-15/16" ± .015			
17	**{Top Chamfer}** 1/16"			
18	**{Inside Diameter}** 2-1/2" ± .015			
19	**{Large Diameter of 7° Taper}** 2-3/8" ± .015			
20	**{Depth of Inside}** 7/8" ± .015			

Project Grade: ____________

Short Diamond Holder

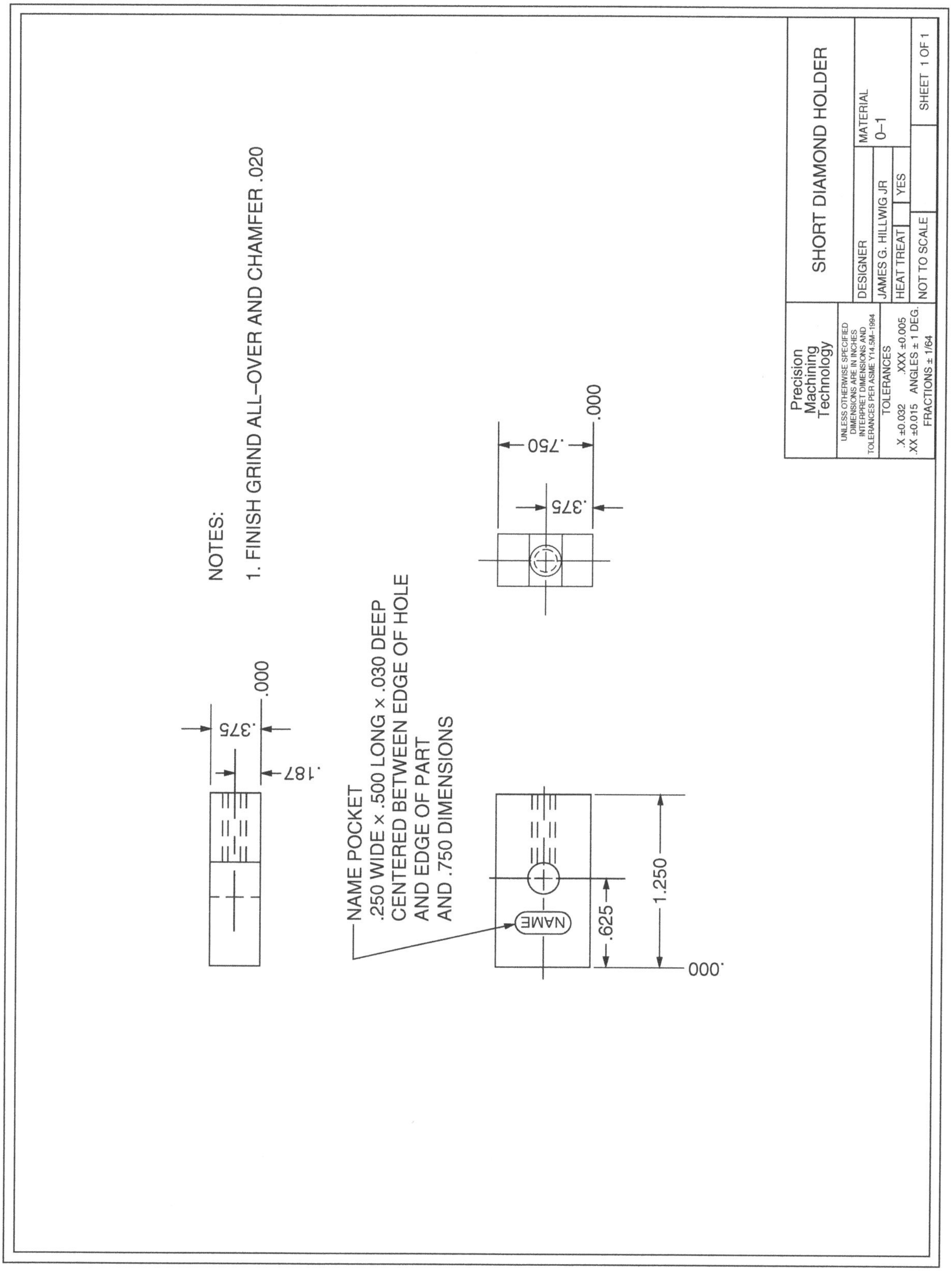

NIMS Skills Practiced

- Job Process Planning
- Layout
- Surface Grinding and Safety
- Vertical Milling and Safety
- Squaring up a Block

Equipment

Power saw, milling machine, surface grinder

Tools Needed

Scribe	6″ steel rule	Layout dye
Dead blow hammer	File	File card
15/64 and #21 drill bits	1/4″ reamer	10-32 tap
90° countersink	Various sizes of endmills	Parallels
Edge finder	Center drill	Vise stop
Machinist square	Grinding wheel	Grinder vise
Tramming items for mill	Stamps	Tap stand

Material Needed

- 3/8″ × 3/4″ × 1-3/8″ tool steel (oversize stock)

Safety

1. Follow all shop safety rules given to you by your instructor and be sure to pass any safety tests given before using shop equipment.
2. Follow all safety rules for power equipment.
3. Be careful of sharp tools when using them.
4. Make sure the file has a handle on it.
5. Be sure all cutting tools are sharp and care is taken while handling them.
6. Be sure ways are locked on mill when necessary.
7. Ring grinding wheels before using them.
8. Make sure magnetic chuck is on when surface grinding.

Order of Operations

1. Look over blueprint to determine tolerances, tools, and material needed.
2. Gather tools needed, figure out RPMs for required tools, and write down their RPMs on sheet.
3. Measure, mark, and cut stock 3/8″ thick × 3/4″ wide; saw 1-3/8″ long.
4. Make sure milling machine is in tram.
5. Deburr all over.
6. Place piece in mill vise on parallels with sawed end sticking out, tap down with dead blow hammer to seat part on parallels, and mill just enough off the end to clean it up.
7. Deburr the end you just milled, flip part 180° on parallels in vise, tap down with dead blow hammer to seat part on parallels, and mill to length.

8. Paint part with layout dye and use height gage to lay out hole locations, being sure to leave grind stock on each end.
9. Place part in vise on parallels against vise stop; tap down with dead blow hammer to seat part on parallels, then set vise stop.
10. Insert edge finder in collet, set RPMs to 1,100, and edge find the X-axis and the Y-axis. Be sure to allow for 1/2 grind stock on the X-axis.
    a. X-axis should read −.1075 (this includes 1/2 of the grind stock).
    b. Y-axis should read .1000.
11. Place center drill in chuck and move to −.375 on Y-axis and .625 on X-axis.
12. Lower center drill to make sure it aligns with layout lines.
13. If location is correct, center drill hole at 1,200 RPMs, change to 15/64 drill, set the RPMs, and drill through part; then insert 1/4 reamer, set RPMs, and ream hole through.
14. Remove part and deburr, rotate part 90°, and place the 3/4″ side against back jaw of vise and 3/8″ edge against vise stop.
15. Place center drill in chuck and move to −.187 on Y-axis and .375 on X-axis.
16. Lower center drill to make sure it aligns with layout lines.
17. If location is correct, center drill hole at 1,200 RPMs, change to #21 drill, set the RPMs, and drill −.850 deep; then insert 90° countersink, set RPMs, and countersink hole.
18. Remove part from the vise and deburr hole.
19. Lay out name pocket, 1/4″ wide × 1/2″ long × .030 deep, centered on 3/4″ face centered from edge to edge of hole—this is opposite side of tapped hole.
20. Using a 1/4″ endmill, set RPMs, locate pocket, and mill to print dimensions. Remove part, deburr, place on anvil, and using steel stamps, stamp your name in pocket.
21. Using tap stand, tap the 10-32 hole.

MAKE SURE MAGNETIC CHUCK IS ON WHEN SURFACE GRINDING!

22. Surface grind all sides square and parallel. Deburr after each side is ground.
23. Surface grind the face to clean it up, deburr, then flip the part to grind bottom.
24. Deburr all edges and place all the way down in grinder vise with ground sides against jaws, leaving 3/4″ edge out of top of vise. Leave 1-3/8″ edge sticking out of side of vise so you can turn vise on its side.
25. Grind top of part to clean up and turn vise on side without removing or disturbing part while in vise just to clean it up.
26. Remove from vise, deburr all edges, and place ground side down on chuck with square blocking on each side; grind side to clean it up.
27. Deburr and place last side up with square blocking on each side and clean up surface.
28. Now check hole locations and grind them into location and part to size. Be sure to deburr and block part when needed.
29. Grind a .020 × 45° chamfer on all edges.
30. Inspect the part to print dimensions; record on inspection sheet and turn in.

How to Determine the RPMs

Tool Being Used	Surface Feet per Minute	Formula	RPM
Edge Finder	50	Given	1,000
#2 Center Drill	50	Given	1,200
#3 Center Drill	50	Given	1,100
Drill 15/64	50	$\frac{4 \times 50}{.235}$	
Drill #21	50	$\frac{4 \times 50}{.125}$	

How to Determine the RPMs for Reamers

Tool Being Used	Surface Feet per Minute	Formula	RPM
Reamer 1/8	50	$\frac{4 \times 50}{.250}$	Divide RPM in half for reamers
90° Countersink	50	$\frac{4 \times 50}{\text{Tool}}$	

Student Name: ______________________________ Date Submitted: ______________

Class: ______________ Total Hours on Job: ______________

	Print Dimensions and Tolerances	Student's Inspection	Instructor's Inspection	Instructor Comments
1	Is the part free of burrs and does it meet finish requirements?			
2	Is the part properly identified?			
3	.375 ± .005			
4	.750 ± .005			
5	1.250 ± .005			
6	1/4 Hole Location .625 ± .005			
7	1/4 Hole Location .375 ± .005			
8	10-32 Tapped Hole .187 ± .005			
9	10-32 Tapped Hole .375 ± .005			
10	Are the hole chamfers to size?			
11				
12				
13				
14				
15				

Project Grade: ______________

Parallels

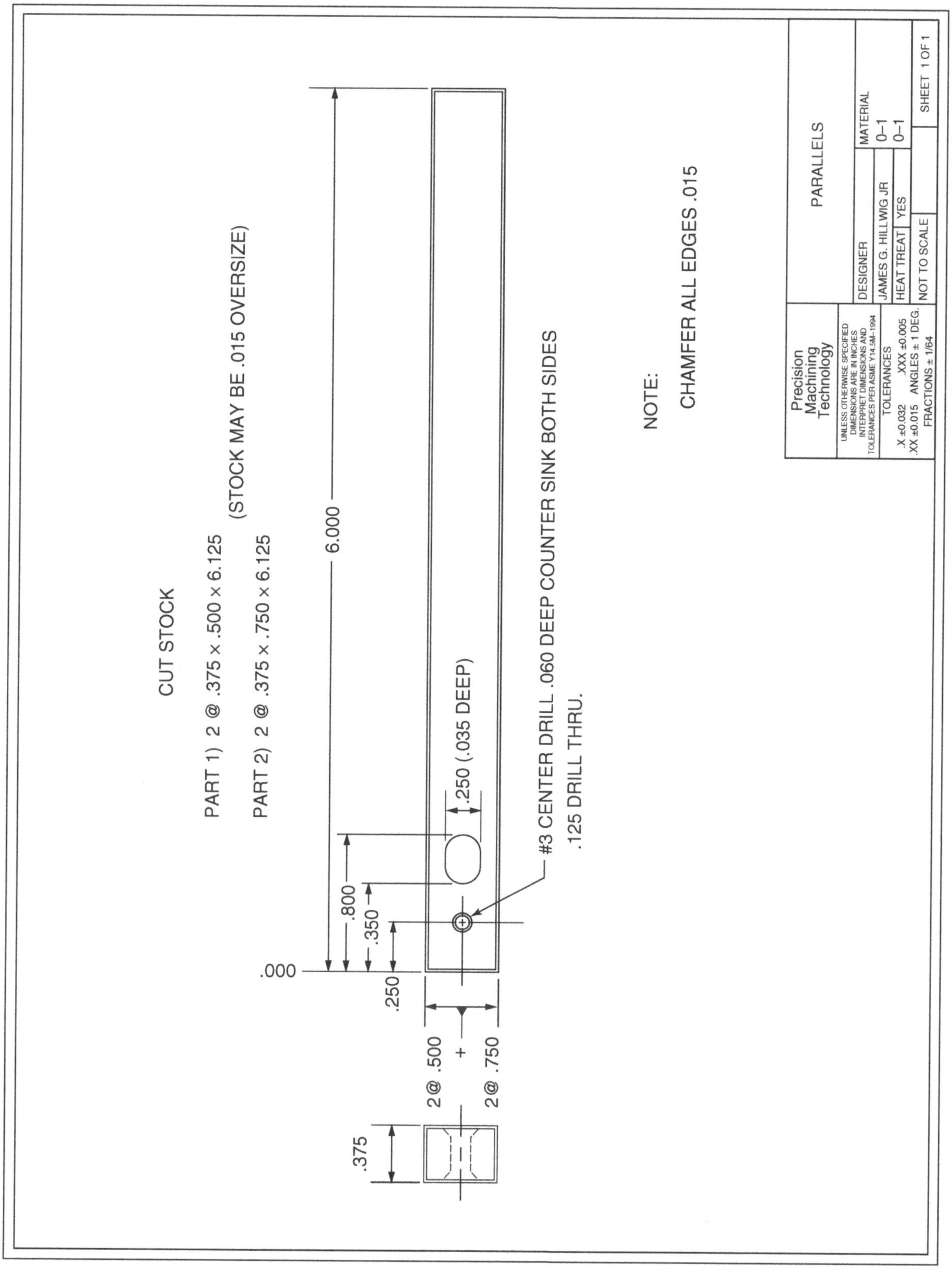
CUT STOCK
PART 1) 2 @ .375 × .500 × 6.125
PART 2) 2 @ .375 × .750 × 6.125
(STOCK MAY BE .015 OVERSIZE)
6.000
.000
.800
.350
.250
.250 (.035 DEEP)
#3 CENTER DRILL .060 DEEP COUNTER SINK BOTH SIDES
.125 DRILL THRU.
.375
2@ .500
2@ .750
NOTE:
CHAMFER ALL EDGES .015
Precision Machining Technology
UNLESS OTHERWISE SPECIFIED DIMENSIONS ARE IN INCHES INTERPRET DIMENSIONS AND TOLERANCES PER ASME Y14.5M-1994
TOLERANCES
.X ±0.032 .XXX ±0.005
.XX ±0.015 ANGLES ± 1 DEG.
FRACTIONS ± 1/64
PARALLELS
DESIGNER
JAMES G. HILLWIG JR
MATERIAL
0–1
0–1
HEAT TREAT
YES
NOT TO SCALE
SHEET 1 OF 1

NIMS Skills Practiced

- Job Process Planning
- Layout
- Surface Grinding and Safety
- Vertical Milling and Safety
- Squaring Up a Block

Equipment

Power saw, milling machine, surface grinder

Tools Needed

Scribe	6″ steel rule	Layout dye
Dead blow hammer	File	File card
1/8″ drill bit	Various sizes of endmills	Parallels
Edge finder	Center drill	Vise stop
Grinding wheel	Grinder vise	Stamps
Tramming items for mill	Large angle plate	Shim stock
Compound clamps		

Material Needed

- 2 @ 3/8″ × 1/2″ × 6-1/8″ tool steel (oversize stock)
- 2 @ 3/8″ × 3/4″ × 6-1/8″ tool steel (oversize stock)

Safety

1. Follow all shop safety rules given to you by your instructor and be sure to pass any safety tests given before using shop equipment.
2. Follow all safety rules for power equipment.
3. Be careful of sharp tools when using them.
4. Make sure the file has a handle on it.
5. Be sure all cutting tools are sharp and care is taken while handling them.
6. Be sure ways are locked on mill when necessary.
7. Ring grinding wheels before using them.
8. Make sure magnetic chuck is on when surface grinding.

Order of Operations

1. Look over blueprint to determine tolerances, tools, and materials needed.
2. Gather tools needed, figure out RPMs for required tools, and write down their RPMs on sheet.
3. Measure, mark, and cut 2 pieces of stock 3/8″ thick × 1/2″ wide; saw 6-1/8″ long.
4. Measure, mark, and cut 2 pieces of stock 3/8″ thick × 3/4″ wide; saw 6-1/8″ long.
5. Make sure milling machine is in tram.
6. Deburr all over.
7. Place piece in mill vise on parallels with sawed end sticking out; tap down with dead blow hammer to seat part on parallels, then mill just enough off the end to clean it up.

8. Do this to all four pieces.
9. Deburr the end you just milled, flip part 180° on parallels in vise, tap down with dead blow hammer to seat part on parallels, and set your vise stop and mill to length.
10. Set zero once the part is to size, then repeat with each part.
11. Place part in vise on parallels against vise stop; tap down with dead blow hammer to seat part on parallels, then set vise stop.
12. Paint part with layout dye and use height gage to lay out hole locations, being sure to leave grind stock on each end.
13. Insert edge finder in collet, set RPMs to 1,100, and edge find the X-axis and the Y-axis. Be sure to allow for 1/2 grind stock on the X-axis.
    a. X-axis should read −.1075 (this includes 1/2 of the grind stock).
    b. Y-axis should read .1075 (this includes 1/2 of the grind stock).
14. Place center drill in chuck and move to hole location.
15. Lower center drill to make sure it aligns with layout lines.
16. If location is correct, center drill hole at 1,200 RPMs, change to 1/8″ drill, set the RPMs, and drill through part. Repeat process on second one of the same size.
17. Then change Y location for second set and center drill and drill each one.
18. Put a 1/4″ endmill in collet and move to the location of the name pocket. Mill the name pocket .030 deep in each parallel.

MAKE SURE MAGNETIC CHUCK IS ON WHEN SURFACE GRINDING!

19. Surface grind all sides square and parallel. Deburr after each side is ground.
20. Surface grind the face to clean it up, deburr, then flip the part to grind bottom.
21. Deburr all edges, then place in grinder vise on parallels with ground sides against jaws.
22. Grind top of part to clean up; repeat for each parallel.
23. Remove from vise, deburr all edges, place ground side down on chuck with square blocking on each side, and grind side to clean it up. Grind them in pairs from this point on.
24. Deburr and place in large angle plate, secure against the inside corner of the angle plate with the compound clamps, and clean up end surface.
25. Turn end for end and repeat.
26. Now check hole locations and grind them into location and part to size. Be sure to deburr and block part when needed.
27. These must be flat and parallel. You might need to use shim stock to remove any bows in the parallels.
28. Grind a .015 × 45° chamfer on all edges.
29. Inspect the part to print dimensions; record on inspection sheet and turn in.

Send to Heat Treat

Student Name: ______________________ Date Submitted: ____________

Class: ____________ Total Hours on Job: ____________

	Print Dimensions and Tolerances	Student's Inspection	Instructor's Inspection	Instructor Comments
1	Is the part free of burrs and does it meet finish requirements?			
2	Is the part properly identified?			
3	6.000 ± .005			
4	.375 ± .005			
5	.500 ± .005			
6	Ø 1/8 Thru Hole ± .015			
7	Location of 1/8 Ø Hole .250 ± .005			
8	Name Slot Length .450 ± .005			
9	Name Slot Width .250 ± .005			
10	Name Slot Depth .035 ± .005			
11	Both Parallels Ground to Match and Flat to ± .0002			
12	6.000 ± .005			
13	.375 ± .005			
14	.750 ± .005			
15	Ø 1/8 thru Hole ± .015			
16	Location of 1/8 Ø Hole .250 ± .005			
17	Name Slot Length .450 ± .005			
18	Name Slot Width .250 ± .005			
19	Name Slot Depth .035 ± .005			
20	Both Parallels Ground to Match and Flat to ± .0002			

Project Grade: ____________

Mill Angles

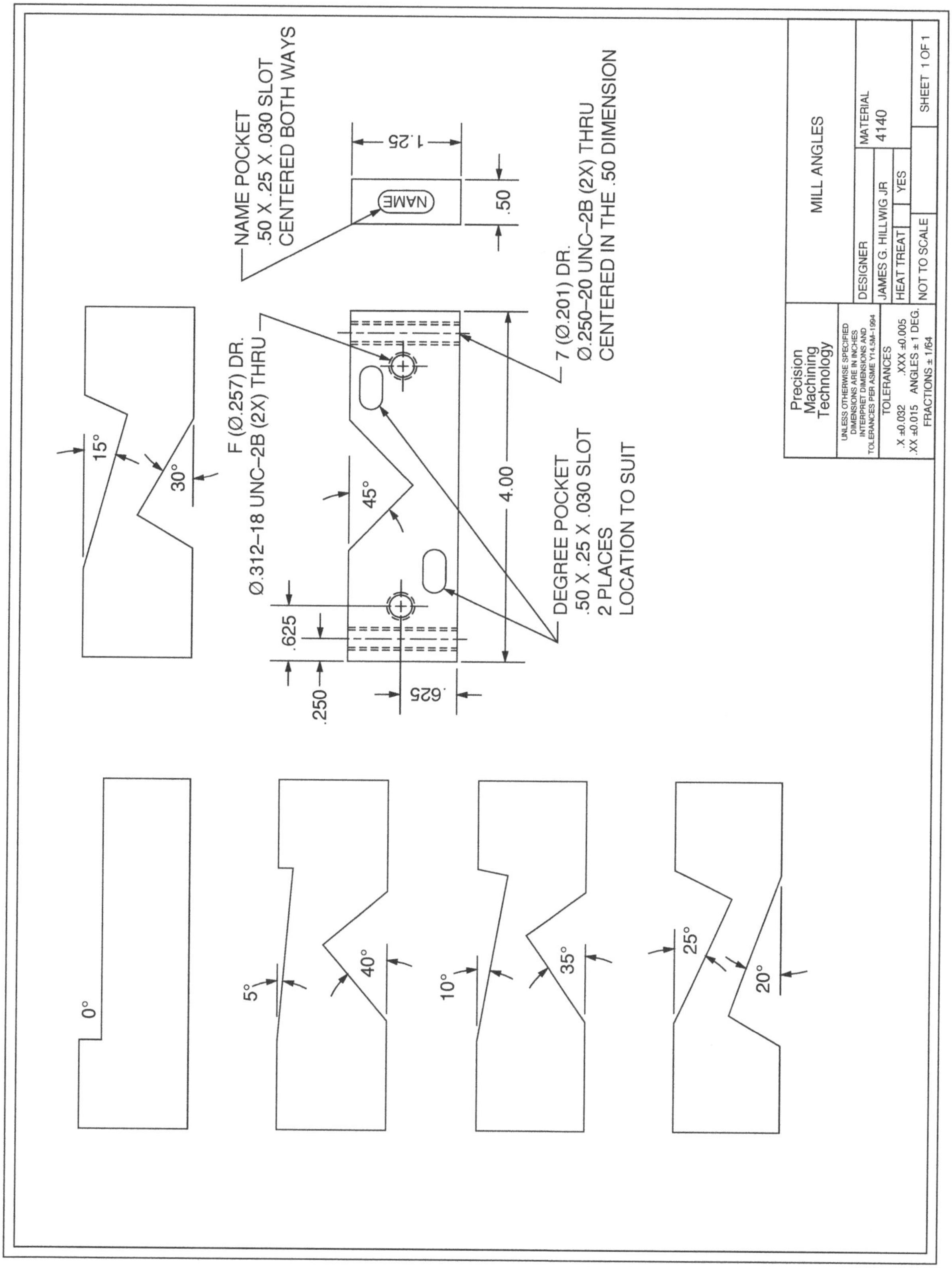
NAME POCKET
.50 X .25 X .030 SLOT
CENTERED BOTH WAYS
1.25
.50
NAME
F (Ø.257) DR.
Ø.312–18 UNC–2B (2X) THRU
7 (Ø.201) DR.
Ø.250–20 UNC–2B (2X) THRU
CENTERED IN THE .50 DIMENSION
4.00
DEGREE POCKET
.50 X .25 X .030 SLOT
2 PLACES
LOCATION TO SUIT
.625
.250
.625
15°
30°
45°
0°
5°
40°
10°
35°
25°
20°
MILL ANGLES
Precision
Machining
Technology
UNLESS OTHERWISE SPECIFIED
DIMENSIONS ARE IN INCHES
INTERPRET DIMENSIONS AND
TOLERANCES PER ASME Y14.5M-1994
TOLERANCES
.X ±0.032 .XXX ±0.005
.XX ±0.015 ANGLES ± 1 DEG.
FRACTIONS ± 1/64
DESIGNER
JAMES G. HILLWIG JR
MATERIAL
4140
HEAT TREAT
YES
NOT TO SCALE
SHEET 1 OF 1

NIMS Skills Practiced

- Job Process Planning
- Layout
- Surface Grinding and Safety
- Vertical Milling and Safety
- Squaring Up a Block

Equipment

Power saw, milling machine, surface grinder

Tools Needed

Scribe
Dead blow hammer
#7 letter "F" drill bits
Edge finder
Grinding wheel
Tramming items for mill
Adjustable mill angle
6" steel rule
File
Various sizes of endmills
Center drill
Grinder vise
1/4 -20 tap
Magnetic sine chuck
Layout dye
File card
Parallels
Vise stop
Stamps
.312-18 tap

Material Needed

- 7 @ .5" × 1.25" × 4-1/8" tool steel (oversize stock)

Safety

1. Follow all shop safety rules given to you by your instructor and be sure to pass any safety tests given before using shop equipment.
2. Follow all safety rules for power equipment.
3. Be careful of sharp tools when using them.
4. Make sure the file has a handle on it.
5. Be sure all cutting tools are sharp and care is taken while handling them.
6. Be sure ways are locked on mill when necessary.
7. Ring grinding wheels before using them.
8. Make sure magnetic chuck is on when surface grinding.

Order of Operations

1. Look over blueprint to determine tolerances, tools, and material needed.
2. Gather tools needed, figure out RPMs for required tools, and write down their RPMs on sheet.
3. Measure, mark, and cut 7 pieces of stock.
4. Make sure milling machine is in tram.
5. Deburr all over.
6. Place piece in mill vise on parallels, square it up, and take it to size.
7. Do this to all 7 pieces.
8. Lay out hole locations, being sure to leave grind stock on each end.
9. Place part in vise and edge find.

10. Place center drill in chuck and center drill holes.
11. Swap blocks and repeat process until all holes are drilled.
12. Now chamfer each hole.
13. Now tap each hole to correct size.
14. Lay out the angles on each part and write the angle for each angle so you don't get confused.
15. Place each part in the vise on the adjustable mill angle and mill each angle to print dimensions.
16. Cut the angle pockets and name pockets in each piece where indicated and stamp them accordingly.

MAKE SURE MAGNETIC CHUCK IS ON WHEN SURFACE GRINDING!

17. Surface grind all sides square and parallel. Deburr after each side is ground.
18. These must be flat, square, and parallel.
19. Using a cup wheel or dished-out wheel, place parts on magnetic sine chuck and grind each of the angles.
20. Grind a .015 × 45° chamfer on all outside edges. Do not chamfer the angles.
21. Inspect the part to print dimensions; record on inspection sheet and turn in.

Send to Heat Treat

Student Name: ______________________ Date Submitted: ______________

Class: ______________ Total Hours on Job: ______________

	Print Dimensions and Tolerances	Student's Inspection	Instructor's Inspection	Instructor Comments
1	Is the part free of burrs and does it meet finish requirements?			
2	Is the part properly identified?			
3	**{Length}** 4. ± .015			
4	**{Thickness}** .50 ± .015			
5	**{Height}** 1.25 ± .015			
6	**{.312-18 Holes}** .625 × .625 ± .005			
7	**{1/4-20 Holes}** .250 × .250 ± .005			
8	**{Angle}** 0° ± 1/2°			
9	**{Angle}** 5° ± 1/2°			
10	**{Angle}** 40° ± 1/2°			
11	**{Angle}** 10° ± 1/2°			
12	**{Angle}** 35° ± 1/2°			
13	**{Angle}** 25° ± 1/2°			
14	**{Angle}** 20° ± 1/2°			
15	**{Angle}** 15° ± 1/2°			
16	**{Angle}** 30° ± 1/2°			
17	**{Angle}** 45° ± 1/2°			
18	**{Angle}** 45° ± 1/2°			

Project Grade: ______________

1-2-3 Block

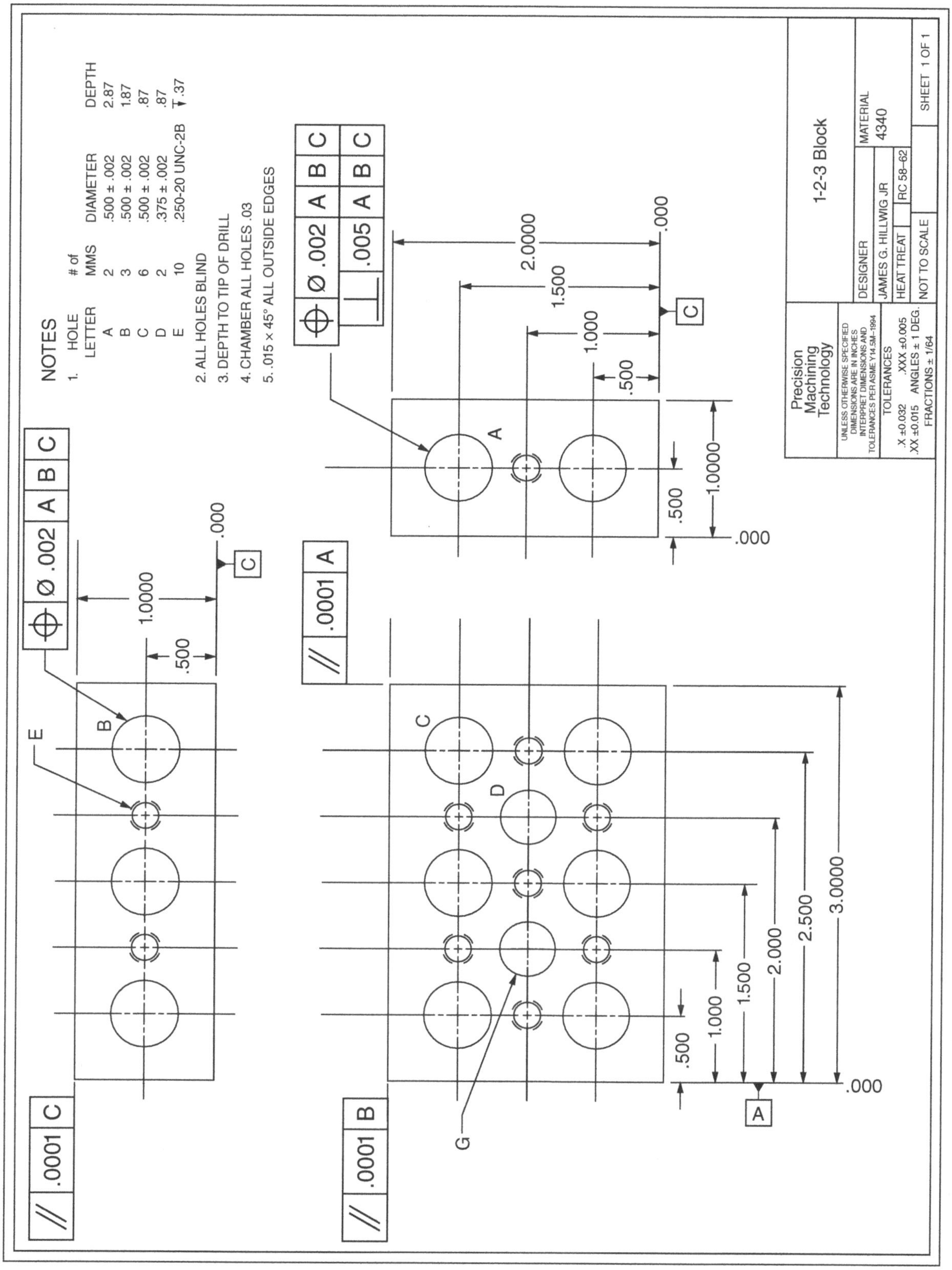
NOTES
1. HOLE LETTER / # of MMS / DIAMETER / DEPTH
A 2 .500 ± .002 2.87
B 3 .500 ± .002 1.87
C 6 .500 ± .002 .87
D 2 .375 ± .002 .87
E 10 .250-20 UNC-2B ↧.37
2. ALL HOLES BLIND
3. DEPTH TO TIP OF DRILL
4. CHAMBER ALL HOLES .03
5. .015 x 45° ALL OUTSIDE EDGES
⌖ Ø .002 A B C
⊥ .005 A B C
// .0001 A
// .0001 B
// .0001 C
A
B
C
D
E
G
.000
.500
1.000
1.500
2.000
2.500
3.0000
2.0000
1.0000
Precision Machining Technology
UNLESS OTHERWISE SPECIFIED DIMENSIONS ARE IN INCHES INTERPRET DIMENSIONS AND TOLERANCES PER ASME Y14.5M-1994
TOLERANCES
.X ±0.032 .XXX ±0.005
.XX ±0.015 ANGLES ± 1 DEG.
FRACTIONS ± 1/64
1-2-3 Block
DESIGNER
JAMES G. HILLWIG JR
MATERIAL
4340
HEAT TREAT
RC 58–62
NOT TO SCALE
SHEET 1 OF 1

NIMS Skills Practiced

- Job Process Planning
- Layout
- Surface Grinding and Safety
- Vertical Milling and Safety
- Squaring Up a Block

Equipment

Power saw, milling machine, surface grinder

Tools Needed

Scribe
Dead blow hammer
Various sizes of drill bits
Edge finder
Grinding wheel
Tramming items for mill
6″ steel rule
File
Various sizes of endmills
Center drill
Grinder vise
Layout dye
File card
Parallels
Vise stop
Stamps

Material Needed

- 2 @ 1″ × 2″ × 3-1/8″ tool steel (oversize stock)

Safety

1. Follow all shop safety rules given to you by your instructor and be sure to pass any safety tests given before using shop equipment.
2. Follow all safety rules for power equipment.
3. Be careful of sharp tools when using them.
4. Make sure the file has a handle on it.
5. Be sure all cutting tools are sharp and care is taken while handling them.
6. Be sure ways are locked on mill when necessary.
7. Ring grinding wheels before using them.
8. Make sure magnetic chuck is on when surface grinding.

Order of Operations

1. Look over blueprint to determine tolerances, tools, and material needed.
2. Gather tools needed, figure out RPMs for required tools, and write down their RPMs on sheet.
3. Measure, mark, and cut 2 pieces of stock 1″ thick × 2″ wide; saw 3-1/8″ long.
4. Make sure milling machine is in tram.
5. Deburr all over.
6. Place piece in mill vise on parallels with sawed end sticking out; tap down with dead blow hammer to seat part on parallels, then mill just enough off the end to clean it up.
7. Do this to both pieces.
8. Deburr the end you just milled, flip part 180° on parallels in vise, tap down with dead blow hammer to seat part on parallels, set your vise stop, and mill to length.

9. Set the zero once the part is to size, then mill next part to same zero.
10. Paint part with layout dye and use height gage to lay out hole locations, being sure to leave grind stock on each end.
11. Place part in vise on parallels against vise stop, tap down with dead blow hammer to seat part vise bottom so you can drill side #1, then set vise stop.
12. Insert edge finder in collet, set RPMs to 1,100, and edge find the X-axis and the Y-axis. Be sure to allow for 1/2 grind stock on the X- and Y-axis.
    a. X-axis should read −.1075 (this includes 1/2 of the grind stock).
    b. Y-axis should read .1075 (this includes 1/2 of the grind stock).
13. Place center drill in chuck and move to first letter "D" hole location.
14. Lower center drill to make sure it aligns with layout lines.
15. If location is correct, center drill hole at 1,200 RPMs.
16. Move to center hole and center drill.
17. Move to third hole and center drill.
18. Swap blocks and repeat process.
19. Now replace center drill with #7 drill and drill each hole again to depths required on print.
20. Now, using a 3/8" drill, drill letter "D" holes to depth in both blocks.
21. Replace with a 1/2" drill, then drill letter "D" holes to depth in both blocks.
22. Now repeat this process on side #2 of both blocks, then side #3, being careful not to break through the bottoms.
23. If you take a 1/2" endmill you can flatten out the bottoms of the three letter "F" holes and stamp your initials inside of the blocks before heat treatment.
24. Now chamfer each hole.
25. Now tap each 1/4-20 hole.

MAKE SURE MAGNETIC CHUCK IS ON WHEN SURFACE GRINDING!

26. Surface grind all sides square and parallel. Deburr after each side is ground.
27. Surface grind face #3 of both blocks together to clean them up, deburr, then flip the parts to grind bottom.
28. Deburr all edges, then place one at a time in grinder vise with ground sides against jaws, leaving side #1 sticking enough out of side of vise that you can place vise on its side to grind end. Grind side #1.
29. Place vise on bottom and grind top of part to clean up side #2.
30. Repeat with second block.
31. Remove from vise, deburr all edges, place ground sides of both blocks down on chuck with square blocking on each side, and grind side to clean it up. Grind them in pairs from this point on.
32. Deburr and place last side up with square blocking on both sides, then clean up end surface.
33. Now check hole locations and grind them into location and part to size. Be sure to deburr and block part when needed.
34. These must be flat, square, and parallel.
35. Grind a .015 × 45° chamfer on all edges.
36. Inspect the part to print dimensions; record on inspection sheet and turn in.

Send to Heat Treat

Student Name: ______________________ Date Submitted: ______________

Class: ______________ Total Hours on Job: ______________

	Print Dimensions and Tolerances	Student's Inspection	Instructor's Inspection	Instructor Comments
1	Is the part free of burrs and does it meet finish requirements?			
2	Is the part properly identified?			
3	**{Length}** 3.000 + .0005 − .0000			
4	**{Width}** 2.000 + .0005 − .0000			
5	**{Height}** 1.000 + .0005 − .0000			
6	**{A}** Ø.500 ± .002			
7	**{A Depth}** 2.87 ± .015			
8	**{B}** Ø .500 ± .002			
9	**{B Depth}** 1.87 ± .015			
10	**{C}** Ø .500 ± .002			
11	**{C Depth}** .87 ± .015			
12	**{D}** Ø .375 ± .002			
13	**{D Depth}** .87 ± .015			
14	**{E}** Ø 1/4 − 20 Tap			
15	**{E Depth}** .37 ± .015			
16	**{Chamfer Outside}** .015 ± .015			
17	**{Holes Chamfered}** .03 ± .015			
18	Hole locations in center?			

Project Grade: ______________

Multiple Diamond Holder

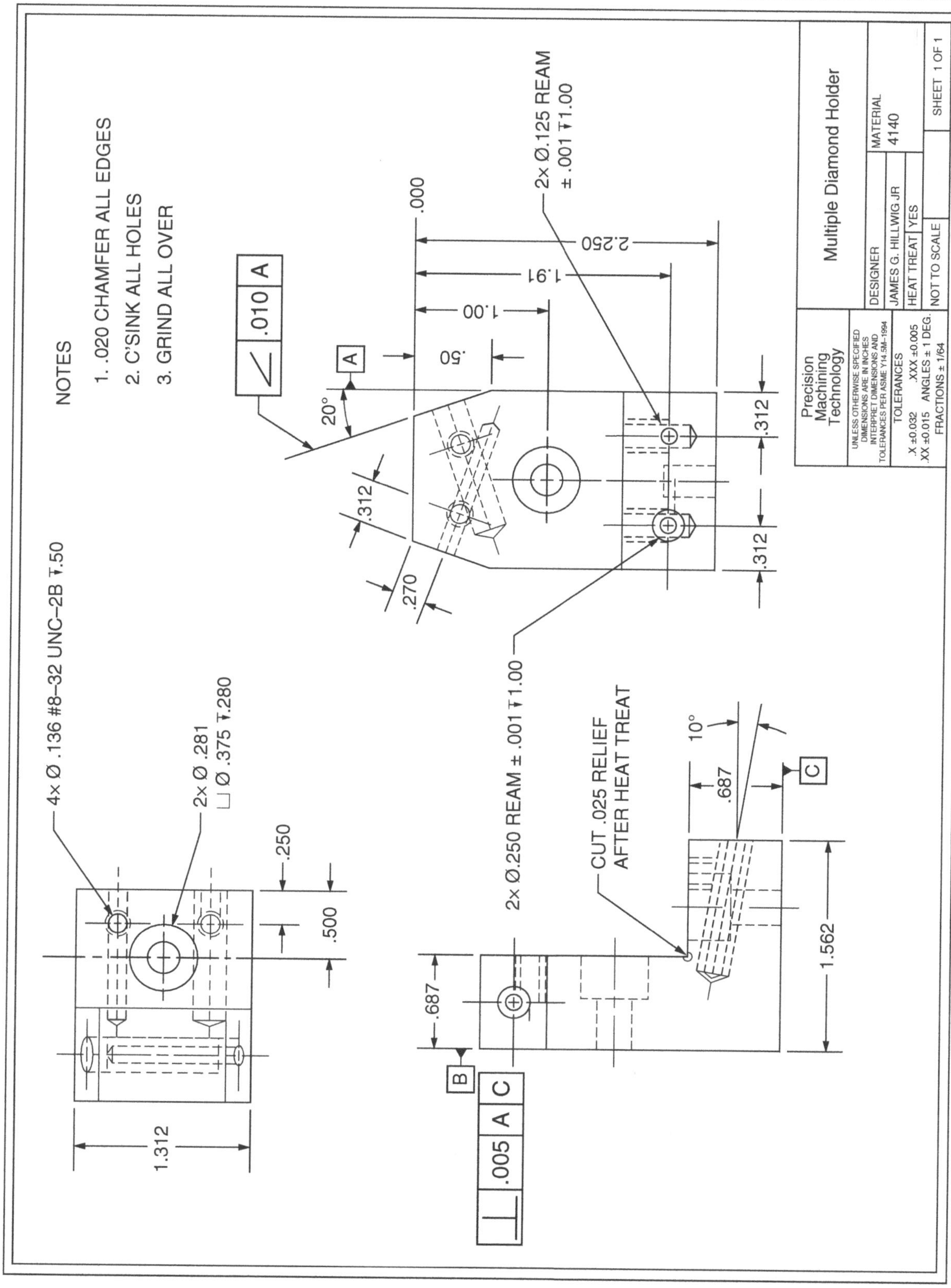

NIMS Skills Practiced

- Job Process Planning
- Layout
- Surface Grinding and Safety
- Vertical Milling and Safety
- Squaring Up a Block

Equipment

Power saw, milling machine, surface grinder

Tools Needed

Scribe
Dead blow hammer
Various sizes of drill bits
Various sizes of reamers
Edge finder
Adjustable mill angle
Tramming items for mill
8-32 tap
6″ steel rule
File
Various sizes of endmills
Parallels
Center drill
Grinding wheel
Magnetic sine plate
Countersink
Layout dye
File card
Stamps
Vise stop
Cup wheel
Grinder vise
Counterbore

Material Needed

- 1.5″ × 1.75″ × 2.375″ tool steel

Safety

1. Follow all shop safety rules given to you by your instructor and be sure to pass any safety tests given before using shop equipment.
2. Follow all safety rules for power equipment.
3. Be careful of sharp tools when using them.
4. Make sure the file has a handle on it.
5. Be sure all cutting tools are sharp and care is taken while handling them.
6. Be sure ways are locked on mill when necessary.
7. Ring grinding wheels before using them.
8. Make sure magnetic chuck is on when surface grinding.

Order of Operations

1. Look over blueprint to determine tolerances, tools, and materials needed.
2. Gather tools needed, figure out RPMs for required tools, and write down their RPMs on sheet.
3. Measure, mark, and cut stock 1.5″ thick × 1.75″ wide; saw 2.375″ long.
4. Make sure milling machine is in tram.
5. Deburr all over.
6. Mill part square.
7. Lay out "L" shape of part.

8. Lay the part on its back in the vise on parallels and mill out the "L" shape, leaving .015" total for grind stock on each .687 dimension.
9. Lay out the rest of the features you need to complete the part.
10. Deburr and place side of part on adjustable mill angle.
11. Mill the 20° angles on part, edge find, then drill the holes in the following order.
12. BE CAREFUL TO DRILL AND REAM THE .125 HOLE FIRST IN THE PART!
13. Drill and ream the .250 hole next.
14. Now repeat the same process for the holes on the 10° surface.
15. Drill, countersink, and carefully tap the 8-32 holes.
16. Drill and counterbore the two holes as shown.
17. Mill a name pocket on the part and stamp your initials in it.

MAKE SURE MAGNETIC CHUCK IS ON WHEN SURFACE GRINDING!

18. Sandblast part to clean it up for grinding.
19. Using the cutoff wheel, grind the relief in the corner as shown.
20. Surface grind all sides square and parallel; deburr after each side is ground.
21. Grind angles using the magnetic sine bar.
22. With your instructor's assistance, place a cup wheel on the surface grinder and grind the inside of the "L."
23. Now check hole locations and grind them into location and part to size. Be sure to deburr and block part when needed.
24. Grind a .015 × 45° chamfer on all edges.
25. You will need the instructor's help to set up the compound angle needed to grind the inside chamfers.
26. Inspect the part to print dimensions; record on inspection sheet and turn in.

Send to Heat Treat

Student Name: ____________________ Date Submitted: ____________

Class: ____________ Total Hours on Job: ____________

	Print Dimensions and Tolerances	Student's Inspection	Instructor's Inspection	Instructor Comments
1	Is the part free of burrs and does it meet finish requirements?			
2	Is the part properly identified?			
3	Front View 1.562 ± .005 length			
4	.687 ± .005 length (top)			
5	.687 ± .005 length (right side)			
6	10° angle (2 holes)			
7	Right Side View 2.250 ± .005 length			
8	1.00 ± .015			
9	.50 length ±.015			
10	1.91 ± .015			
11	.312 ± .005 length (left side)			
12	.312 ± .005 length (right side)			
13	20° angle (2 places)			
14	# 8-32 THD. .500 DP .360 × .382 loc.			
15	# 8-32 THD. .500 DP .360 × .930 loc.			
16	Ø.125 ± .005 1.00 DP (2 holes)			
17	Ø.250 ± .005 1.00 DP (2 holes)			
18	Counterbore for .250 Allen Screw 1.00 × .656 loc.			
19	Top View 1.312 ± .005 length			
20	Ø.250 ± .005 1.00 DP			
21	Ø.125 ± .005 1.00 DP			
22	Counterbore for .250 Allen Screw .500 × .656 loc.			
23	# 8-32 THD. .500 DP .250 ×.312 loc. (top hole)			
24	# 8-32 THD. .500 DP .250 × .312 loc. (bottom hole)			
25	.020 Chamfer All Edges			

Project Grade: ____________

Male/Female Tap Center

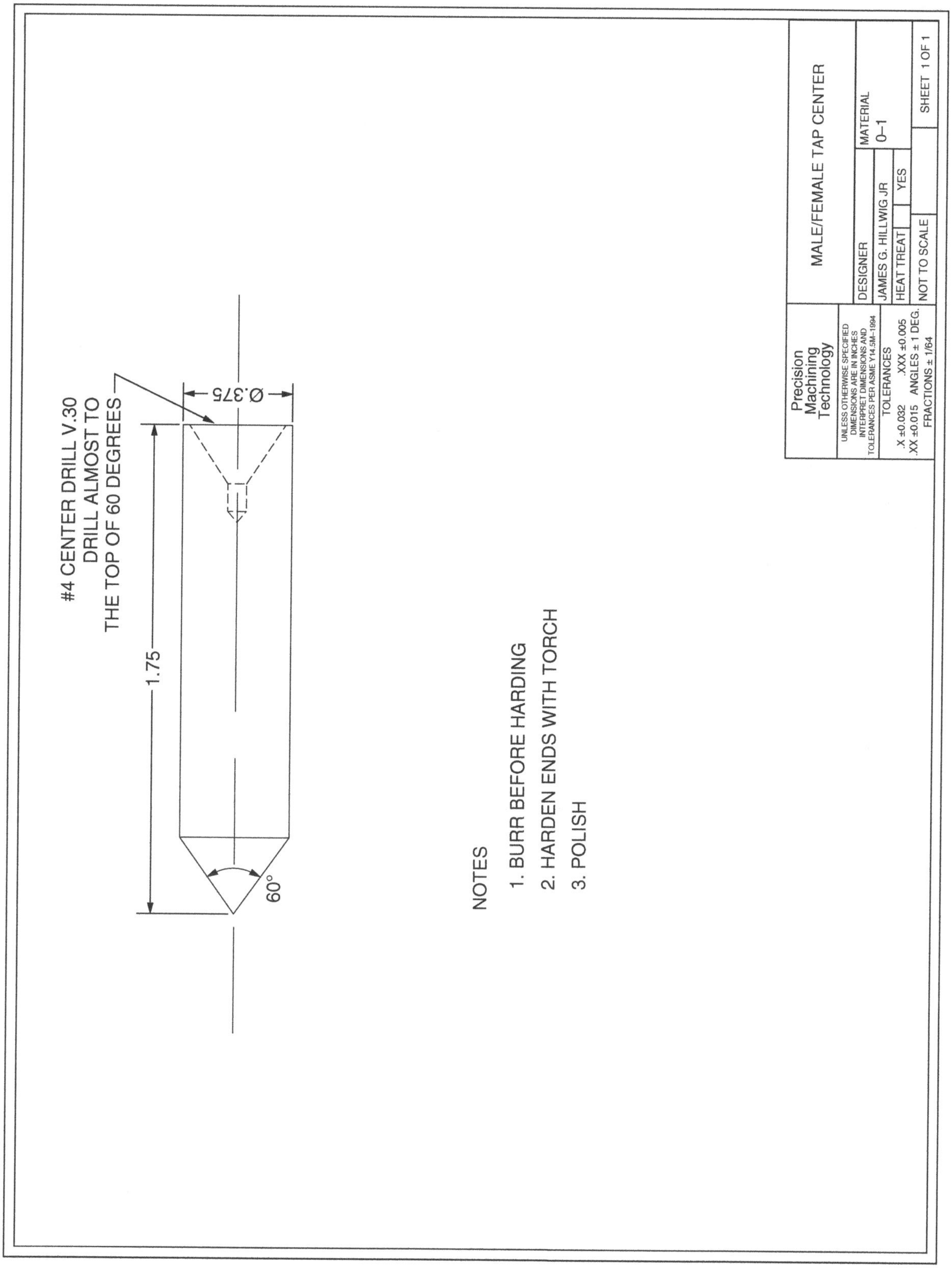

NIMS Skills Practiced

- Job Process Planning
- Turning Operations: Chucking

Equipment

Power saw, lathe, bench grinder

Tools Needed

Scribe
File card
Turning, facing tools
6″ steel rule
Various files
Layout dye
Center drill

Material Needed

- 0-1 tool steel 3/8″ × 1-5/8″

Safety

1. Follow all shop safety rules given to you by your instructor and be sure to pass any safety tests given before using shop equipment.
2. Follow all safety rules for power equipment.
3. Be careful of sharp tools when using them.
4. Make sure the files have a handle on them.
5. Be sure all cutting tools are sharp and care is taken while handling them.
6. Be sure chuck is tightened and chuck wrench is never left in chuck.
7. Make sure stock is supported with center when necessary.

Order of Operations

1. Look over blueprint to determine tolerances, tools, and material needed.
2. Gather tools needed, figure out RPMs for required stock diameters and drills used, and write down their RPMs on sheet.
3. Measure and cut stock sizes needed.
4. Place stock in chuck.
5. Face off one end and center drill.
6. Turn part in chuck and face to length.
7. Set the compound rest to 30° and set tool.
8. Using a steady feed with the compound rest, turn until angle comes to a point.
9. Polish taper and rest of part.
10. Inspect part and turn in.
11. Heat-treat part with torches and re-polish.

Student Name: ______________________ Date Submitted: ____________

Class: ______________ Total Hours on Job: __________

	Print Dimensions and Tolerances	Student's Inspection	Instructor's Inspection	Instructor Comments
1	Is the part free of burrs and does it meet finish requirements?			
2	Is the part properly identified?			
3	Ø.375 ± .005			
4	Length 1.75 ± .015			
5	Is the #4 center-drilled hole good?			
6	Is the 60-degree angle good?			
7				
8				
9				
10				
11				
12				
13				
14				
15				

Project Grade: ______________

Lathe Chuck Center

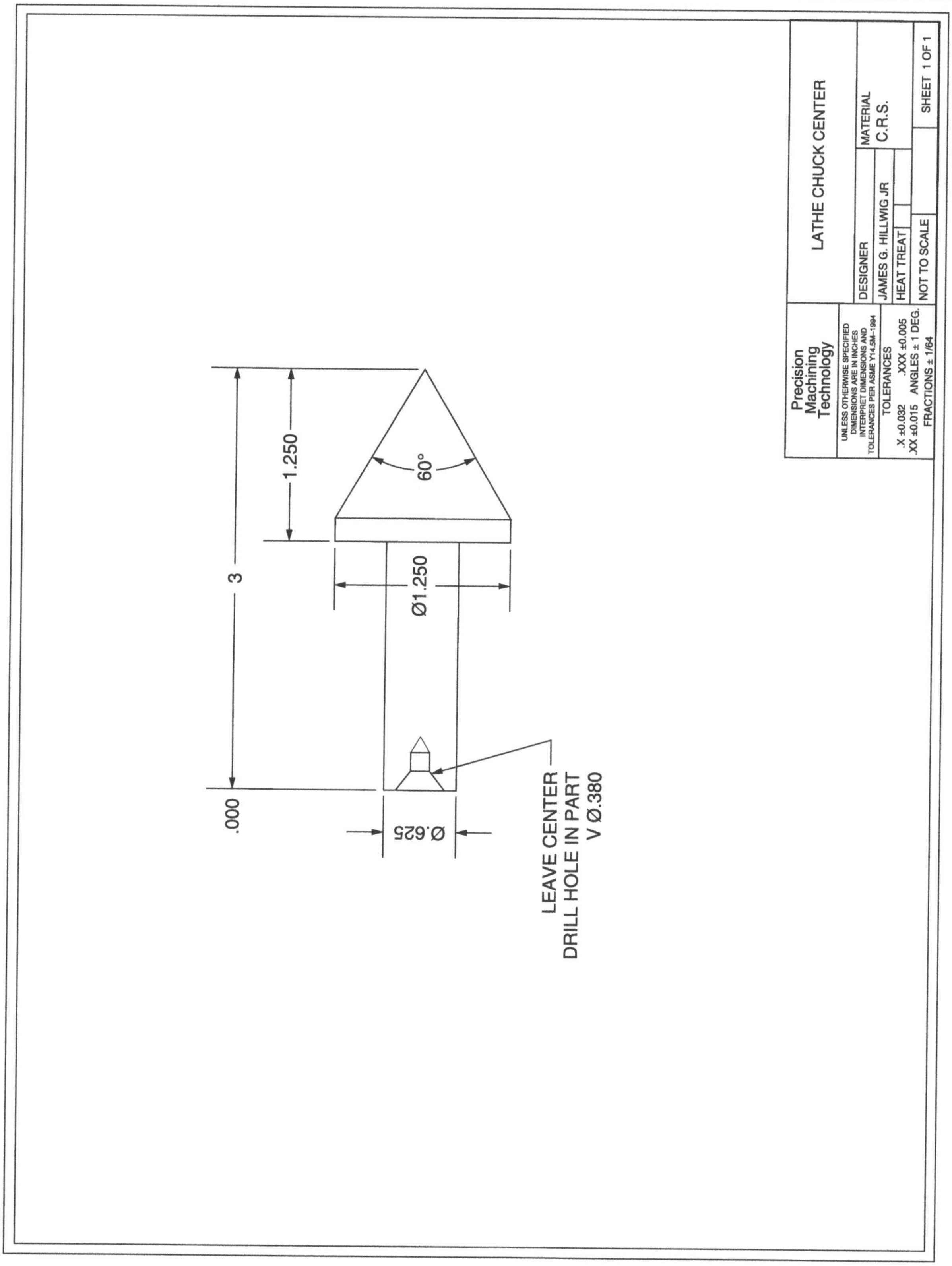

NIMS Skills Practiced

- Job Process Planning
- Turning Operations: Chucking

Equipment

Power saw, lathe, bench grinder

Tools Needed

6″ steel rule
Center drill
File card
Various types of turning and facing tools
Various files

Material Needed

- 1-1/4″ diameter × 3-1/8″ long CRS

Safety

1. Follow all shop safety rules given to you by your instructor and be sure to pass any safety tests given before using shop equipment.
2. Follow all safety rules for power equipment.
3. Be careful of sharp tools when using them.
4. Make sure the files have a handle on them.
5. Be sure all cutting tools are sharp and care is taken while handling them.
6. Be sure chuck is tightened and chuck wrench is never left in chuck.
7. Make sure stock is supported with center when necessary.

Order of Operations

1. Look over blueprint to determine tolerances, tools, and materials needed.
2. Gather tools needed, figure out RPMs for required stock diameters and drills used, and write down their RPMs on sheet.
3. Measure and cut stock sizes needed.
4. Place in chuck and face and center drill one end.
5. Slide part 2″ out of chuck and place tailstock's live center in center-drilled hole.
6. Turn the .501 diameter, leaving a few thousands of an inch to polish to size.
7. Polish to size and finish.
8. Turn part end for end and machine to length.
9. Set up compound rest and turn the 60° included angle.
10. Do not remove or loosen the part from the chuck until angle is complete; if you do, you will need to re-cut the 60° angle.
11. Polish the angle.
12. Fill out inspection sheet and turn in.

Student Name: ______________________ Date Submitted: ____________

Class: ______________ Total Hours on Job: __________

	Print Dimensions and Tolerances	Student's Inspection	Instructor's Inspection	Instructor Comments
1	Is the part free of burrs and does it meet finish requirements?			
2	Is the part properly identified?			
3	3" ± .015			
4	Ø 1.250 ± .005			
5	1.250 ± .005			
6	Ø.625 ± .005			
7	1.750 ± .005			
8	60° Angle Polished			
9	Dimension Polished Ø.625			
10				
11				
12				
13				
14				
15				

Project Grade: ______________

Punch Set

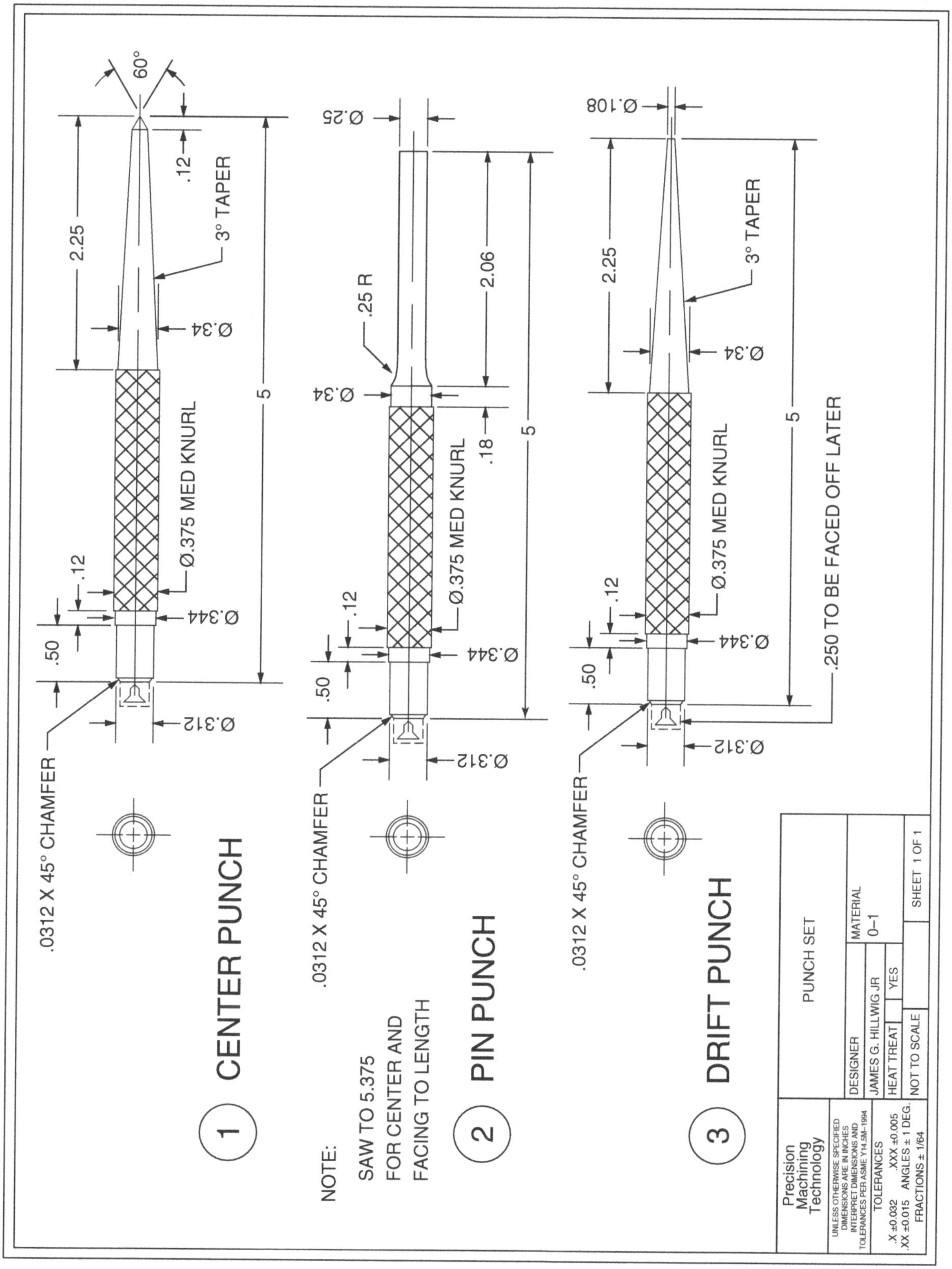
1 CENTER PUNCH
60°
.12
3° TAPER
2.25
Ø.34
5
Ø.375 MED KNURL
.12
Ø.344
.50
Ø.312
.0312 X 45° CHAMFER
2 PIN PUNCH
Ø.25
2.06
.25 R
Ø.34
.18
5
Ø.375 MED KNURL
.12
Ø.344
.50
Ø.312
.0312 X 45° CHAMFER
NOTE:
SAW TO 5.375
FOR CENTER AND
FACING TO LENGTH
3 DRIFT PUNCH
Ø.108
2.25
3° TAPER
Ø.34
5
Ø.375 MED KNURL
.12
Ø.344
.50
.250 TO BE FACED OFF LATER
Ø.312
.0312 X 45° CHAMFER
Precision Machining Technology
UNLESS OTHERWISE SPECIFIED DIMENSIONS ARE IN INCHES INTERPRET DIMENSIONS AND TOLERANCES PER ASME Y14.5M-1994
TOLERANCES
.X ±0.032 .XXX ±0.005
.XX ±0.015 ANGLES ± 1 DEG.
FRACTIONS ± 1/64
PUNCH SET
DESIGNER
JAMES G. HILLWIG JR
MATERIAL
0–1
HEAT TREAT
YES
NOT TO SCALE
SHEET 1 OF 1

NIMS Skills Practiced

- Job Process Planning
- Turning Operations: Chucking

Equipment

Power saw, lathe, bench grinder

Tools Needed

Marker
Various files
Various grits of sandpaper
Radius gages
6″ steel rule
Center drill
Turning and facing tools
File card
Knurling tool
Radius tool

Material Needed

- 3/8″ diameter × 5-3/8″ long

Safety

1. Follow all shop safety rules given to you by your instructor and be sure to pass any safety tests given before using shop equipment.
2. Follow all safety rules for power equipment.
3. Be careful of sharp tools when using them.
4. Make sure the files have a handle on them.
5. Be sure all cutting tools are sharp and care is taken while handling them.
6. Be sure chuck is tightened and chuck wrench is never left in chuck.
7. Make sure stock is supported with center when necessary.

Order of Operations

1. Look over blueprint to determine tolerances, tools, and material needed.
2. Gather tools needed, figure out RPMs for required stock diameters and drills used, and write down their RPMs on sheet.
3. Measure and cut stock sizes needed.
4. Place in chuck and face and center drill one end no deeper than 1/4″ deep. This will be used for knurling only and removed afterward. Do this on each part.
5. Now slide part out of chuck, leaving approx. 1-1/2″ in chuck, placing the tailstock in the center-drilled hole.
6. Using the medium knurl found on the knurling tool, knurl the .375 diameter from the chuck toward the tailstock. Make sure everything is tight.
7. Repeat on the other two parts.
8. Face out the center-drilled hole of each part just enough to remove it, then flip parts end for end and take them to size, 5″.
9. Place parts, one at a time, back in chuck with knurled side sticking out 1″; protect the knurls from being crushed in the chuck.
10. Turn the 11/32 diameter back 5/8″ on each part, leaving a few thousandths to polish.

11. Now turn the .312 diameter 1/2″ on each part, leaving a few thousandths to polish.
12. Using a file, put the chamfer on each part.
13. Flip parts end for end and turn the 11/32 diameter back on each punch 2-1/4″, leaving a few thousandths for polishing.
14. Set the taper on the compound rest at 3° and turn the taper on the center punch and the drift punch; leave a few thousandths for polishing.
15. Using a file, file the 60° angle on the end of the center punch.
16. Place the pin punch in the chuck and turn the .250 diameter back 2-1/16″ minus 1/4″ (to leave stock for the radius). Leave a few thousandths on the .250 diameter for polishing.
17. Place the 1/4″ radius tool in holder and blend the radius in.
18. Polish all the turned surfaces to remove any marks or scratches before heat treatment or they will not come out easily.

Send to Heat Treat

19. Polish all the turned surfaces, leaving the black in the knurls, then dip them in oil and the black will contrast nicely.
20. Wipe off excess oil; fill out inspection sheet and turn in.

Student Name: ______________________ Date Submitted: ____________

Class: ______________ Total Hours on Job: ____________

	Print Dimensions and Tolerances	Student's Inspection	Instructor's Inspection	Instructor Comments
1	Is the part free of burrs and does it meet finish requirements?			
2	Is the part properly identified?			
3	**(Center Punch)** .5 ± .015			
4	(Diameter) Ø.312 ± .005			
5	(Length of Ø.312) .50 ± .015			
6	(Diameter) Ø.344 ± .005			
7	(Length of Ø.344) .12 ± .015			
8	(Diameter/Tapered End) Ø.34 ± .015			
9	(Length of Tapered End) 2.25 ± .015			
10	3° Taper ± 1°			
11	60° Taper ± 1°			
12	45° Chamfer by .0312 ± .005			
13	(Taper) .12 ± .015			
14	**(Pin Punch)** .5 ± .015			
15	(Diameter) Ø.312 ± .005			
16	(Length of Ø.312) .50 ± .015			
17	(Diameter) Ø.344 ± .005			
18	(Length of Ø.344) .12 ± .015			
19	(Diameter .250 end) Ø.34 ± .015			
20	(Length of Ø.34) .18 ± .015			
21	(Length of .250 end) 2.06 ± .015			

(Continued)

	Print Dimensions and Tolerances	Student's Inspection	Instructor's Inspection	Instructor Comments
22	(Diameter of .250 end) Ø.25 ± .015			
23	.25 R ± .015			
24	45° Chamfer by .0312			
25	**(Drift Punch)** 5. ± .015			
26	(Diameter) Ø.312 ± .005			
27	(Length of Ø.312) .50 ± .015			
28	(Diameter) Ø.344 ± .005			
29	(Length of Ø.344) .12 ± .015			
30	(Diameter/Tapered End) Ø.34 ± .015			
31	(Length of Tapered End) 2.25 ± .015			
32	3° Taper ± 1°			
33	(Tapered End) Ø.108 Diameter ± .005			
34	45° Chamfer by .0312 ± .005			

Project Grade: ______________________

Reversible Pocket Scribe

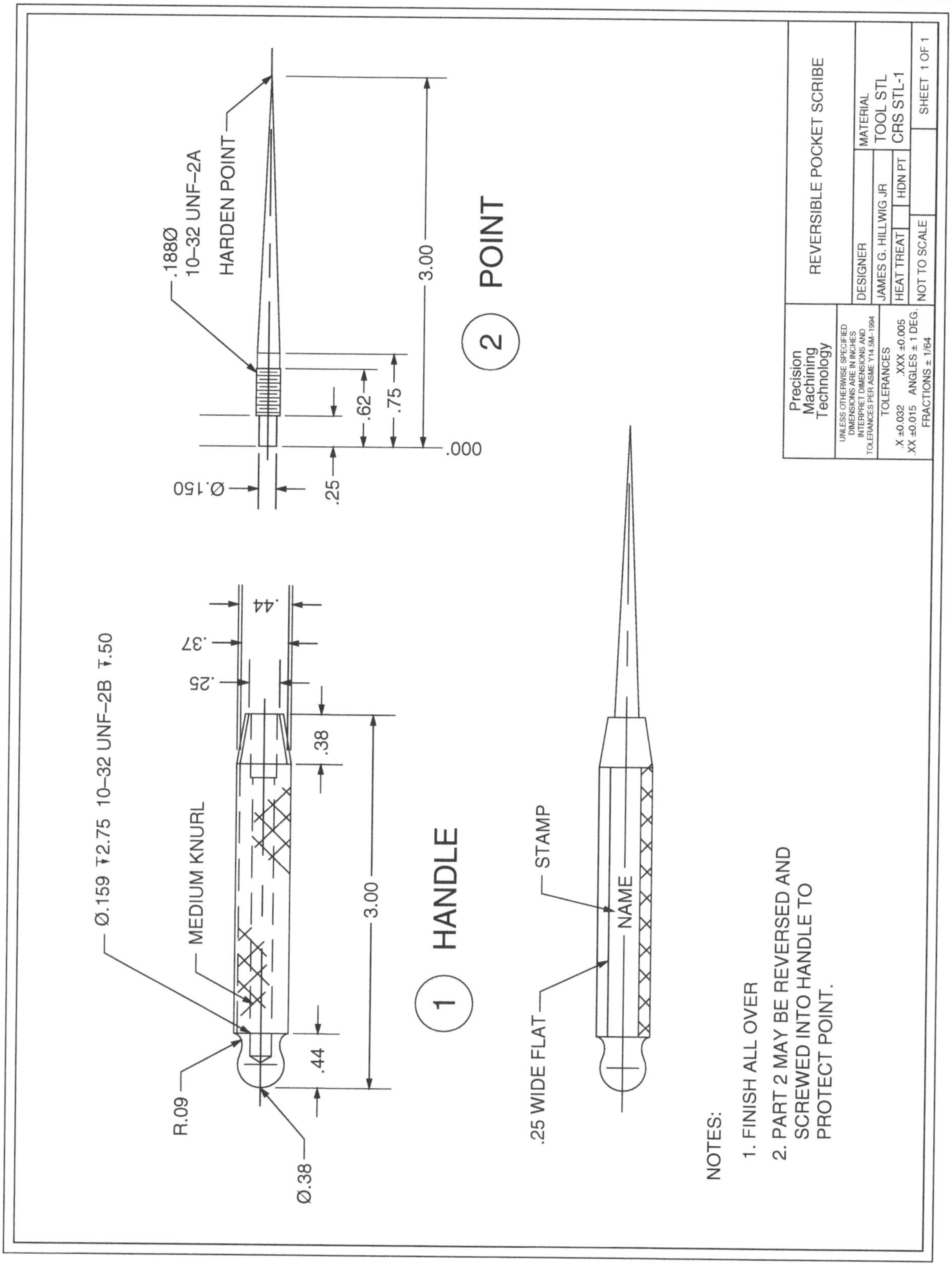

NIMS Skills Practiced

- Job Process Planning
- Turning Operations: Chucking

Equipment

Power saw, lathe, bench grinder

Tools Needed

Marker
Various files
#21 drill (short and long)
Various types of turning and facing tools
6″ steel rule
Center drill
10-31 tap and die
Radius gages
File card
Knurling tool
Various grits of sandpaper

Material Needed

- 1/2″ diameter × 4″ long CRS or aluminum (handle)
- 3/16″ × 3″ long tool steel (Point: this needs to be heat treated 0-1; can be done with torches)

Safety

1. Follow all shop safety rules given to you by your instructor and be sure to pass any safety tests given before using shop equipment.
2. Follow all safety rules for power equipment.
3. Be careful of sharp tools when using them.
4. Make sure the files have a handle on them.
5. Be sure all cutting tools are sharp and care is taken while handling them.
6. Be sure chuck is tightened and chuck wrench is never left in chuck.
7. Make sure stock is supported with center when necessary.

Order of Operations

HANDLE

1. Look over blueprint to determine tolerances, tools, and materials needed.
2. Gather tools needed, figure out RPMs for required stock diameters and drills used, and write down their RPMs on sheet.
3. Measure and cut stock sizes needed.
4. Place in chuck and face and center drill one end; make sure the center-drilled hole is no larger than 3/16″ in diameter. This will be the end that the taper goes on and the #21 drilled hole is in.
5. Turn part end for end (stick center-drilled end into chuck); face and turn to a .400 diameter by .375 long.
6. This step is only done to create a way to hold the part in the chuck so you can safely knurl the part. (It will be cut off later.)
7. Place the .400 diameter you just turned into the chuck. Place the lathe center in the tailstock; place the point of the lathe center into the center-drilled hole you made earlier, which was no larger than 3/16″. Now turn the part to .437 diameter back 3-1/2″ long. This will become the knurled surface.

8. Using the medium knurl found on the knurling tool, knurl the .437 diameter you just turned as far as you safely can without hitting the chuck. Remember, looking at the print, you'll see one end of the handle has a 3/8″ long taper and the other end has a 3/8″ long ball, so don't worry if the knurl doesn't reach both edges of the part.
9. Leaving the part setup as is, remove the knurling tool and put your cutting tool back in. Turn down the end with the center-drilled hole in it, leaving the lathe center in place, to a .375 diameter × .375 long. This is the end the taper will go on.
10. Using simple trig, trig out the angle you need to set the compound rest. See instructor for formula and proper setup of compound rest. Using the handle on the compound rest, turn the taper, leaving a few thousandths for filing and polishing.
11. Turn part end for end, holding the knurls in the jaws (place something on knurls to protect them; see your instructor); face part to 3″ long. You will be removing the .400 diameter × .375 end used to hold the part for knurling.
12. Once the part is 3″ long, turn .375 diameter back 7/16″ long on the same end; this will become the ball on the handle.
13. Use various files to form the ball. File the ball on the part using radius gages to check it. Leave a little for polishing.
14. Turn part end for end, placing ball end in jaws and holding part by knurls (use same protector on the knurls); using the drill chuck in the tailstock, drill with a #21 drill 3/4 of the way up the flutes with a regular-length drill bit. Then see instructor for the extra-long #21 drill bit to drill hole to depth.
15. Tap the hole to print; ask instructor for help to do it on the machine.
16. Polish the taper and ball.
17. You can grind the flat on the handle (using the surface grinder and vise with parallels) according to print, and then stamp your name on the flat if required by instructor.

POINT

1. Use 3/16″ tool steel.
2. File .150 diameter by .250 long on one end.
3. Turn end for end in chuck, leaving about 1″ sticking out to file to the tapered point as shown on print.
4. Keep sliding out a little at a time until .375 of the original 3/16″ remains (this is for the threads).
5. Compare your piece to the print—it should look identical.
6. Now remove the point and hold in vise with soft jaws and thread according to print.
7. Polish.
8. Heat-treat.
9. Polish.

Student Name: ______________________________ Date Submitted: ______________

Class: ______________ Total Hours on Job: ______________

	Print Dimensions and Tolerances	Student's Inspection	Instructor's Inspection	Instructor Comments
1	Is the part free of burrs and does it meet finish requirements?			
2	Is the part properly identified?			
3	**(Handle)** 3.00 ± .015			
4	(Ball) Ø.38 Diameter ± .015			
5	(Ball) .09 Radius ± .015			
6	(Ball) .44 ± .015			
7	(Knurl) Ø.44 ± .015			
8	(Taper) Ø.37 ± .015			
9	(Taper) Ø.25 ± .015			
10	(Taper) .38 ± .015			
11	Is the #21 drill 2.75 deep with a 10-32 thread .50 deep?			
12	**(Point)** 3.00 ± .015			
13	(Point) Ø.150 Diameter ± .005			
14	(Point) .150 Diameter .25 long ± .015			
15	(10-32 Thread) Ø.188 Diameter ± .000-.003			
16	(10-32 Thread) .38 long ± .015			
17	Is .25 wide flat ground to print?			
18	Is the point polished properly?			
19	Is the handle polished properly?			
20	Does the point fit properly into the handle?			

Project Grade: ______________

Small Brass Hammer

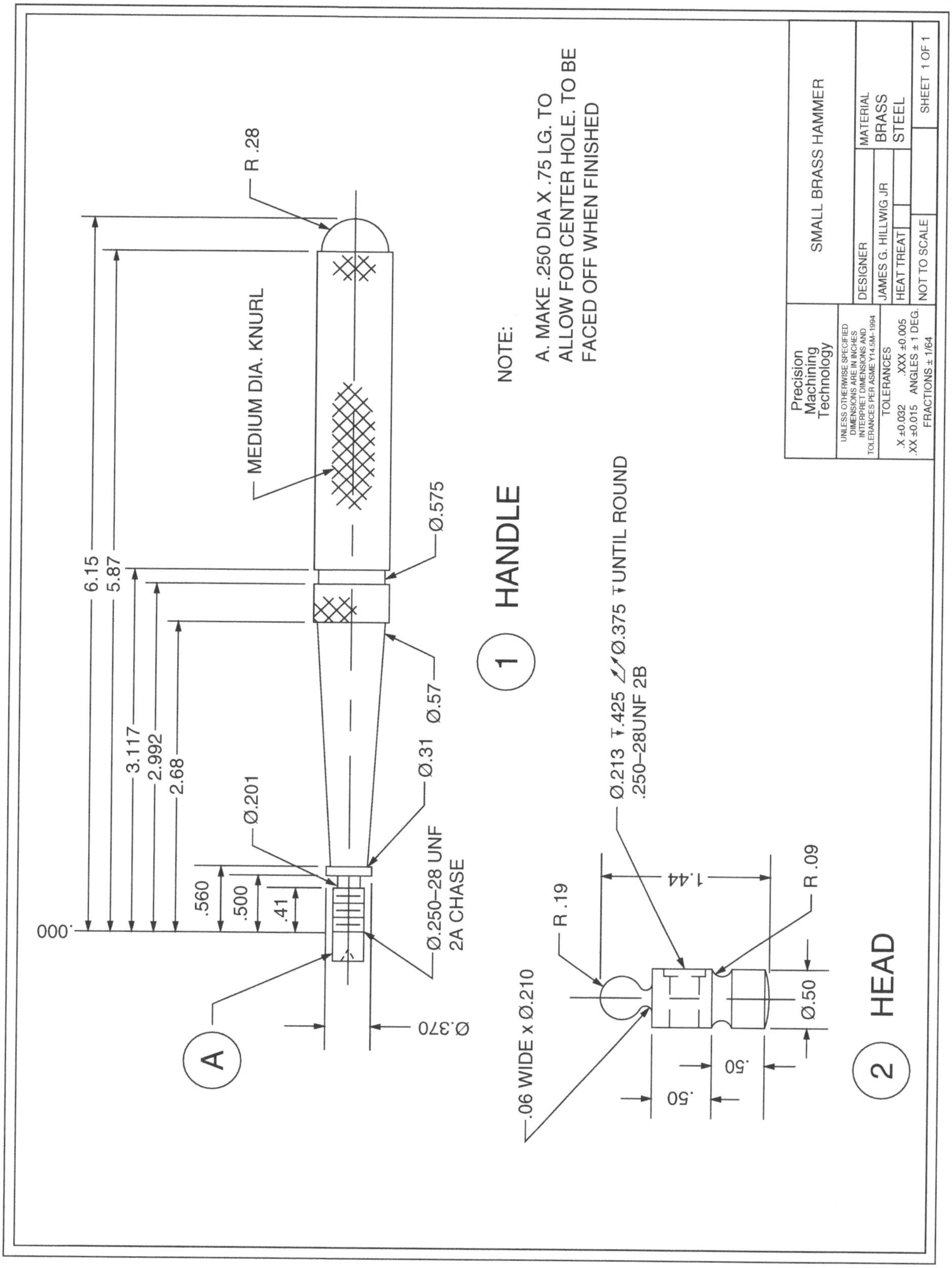

NIMS Skills Practiced

- Job Process Planning
- Turning Operations: Chucking
- Milling: Drilling and Tapping

Equipment

Power saw, lathe, bench grinder, milling machine

Tools Needed

Marker
Various files
#3 drill
Various types of turning and facing tools
6″ steel rule
Center drill
1/4-28 tap
Radius gages
1/8″ stamps
File card
Knurling tool
Various grits of sandpaper
3/8″ endmill
Hammer

Material Needed

- 5/8″ diameter × 6-3/4″ long CRS or aluminum (handle)
- 1/2″ diameter × 1-1/2″ long brass

Safety

1. Follow all shop safety rules given to you by your instructor and be sure to pass any safety tests given before using shop equipment.
2. Follow all safety rules for power equipment.
3. Be careful of sharp tools when using them.
4. Make sure the files have a handle on them.
5. Be sure all cutting tools are sharp and care is taken while handling them.
6. Be sure chuck is tightened and chuck wrench is never left in chuck.
7. Make sure stock is supported with center when necessary.

Order of Operations

HANDLE

1. Saw 5/8″ Ø CRS (mild steel) 7-1/4″ long.
2. Face and center drill both ends. Just face enough to get ends cleaned up.
   a. On the 1/4″ – 28 end, center drill NO larger than 3/16″ Ø.
3. Medium knurl 5-1/2″ back.
4. On the end NOT knurled, turn .370 Ø × .850 back.
5. Turn .250 Ø × .750 back.
6. Groove .201 Ø × .090 wide as shown.
7. Thread 1/4 –28 end.
8. Face handle to length, removing center drill on radius end.
9. Turn .562 Ø back .281.
   a. The .281 radius doubled equals .562 Ø.

10. Turn .575 Ø back 2.180 from the .370 Ø shoulders.
11. Trig out angle and set compound rest to this angle.
    a. Double-check angle (trig) to be sure it is correct.
12. Cut taper both directions with proper turning tools.
    a. Be careful not to cut off the .370 Ø shoulder.
13. Groove 1/8" wide ring on tapered end, deep enough to remove knurl, leaving 5/16" knurl as shown.
14. Polish the taper and the radius.
15. Stamp initials in the 1/8"-wide ring using 1/8" letters.

HEAD

1. Chuck 1/2" Ø brass bar about a foot long.
2. Face-off, then turn 3/8" Ø × 7/16" back.
3. Groove 1/16" radius to .250 Ø with radius tool.
4. Use file to form ball.
5. Groove .09 radius to .09 deep each side with radius tool.
6. Polish, then cut off head using cutoff tool to 1-1/2" long and face to 1.450 in length.
7. File convex radius on end.
    a. Watch 1.44 length.

****ONCE YOU PUT THE PART IN THE MILL, DO NOT REMOVE UNTIL STEP #11 IS COMPLETELY FINISHED!**

8. On milling machine, center drill and drill with a #3 drill .430 deep.
    a. Do not break through bottom.
9. Tap 1/4 – 28 approx. 1/4" deep.
    a. Do not break through bottom.
10. Spot-face with 3/8" Ø endmill, until a full 3/8" Ø circle is achieved.
11. Remove from vise and re-tap by hand with a plug tap and then with a bottom tap.
12. Saw off or grind off 1/4" – 28 threads on handle until the shoulder of handle fits brass head tightly at the spot-face.
13. Fill out evaluation sheet and turn part and evaluation in to the instructor for final evaluation and grade.

Student Name: ______________________ Date Submitted: ____________

Class: ______________ Total Hours on Job: ____________

	Print Dimensions and Tolerances	Student's Inspection	Instructor's Inspection	Instructor Comments
1	Is the part free of burrs and does it meet finish requirements?			
2	Is the part properly identified?			
3	**{Handle}** 6.15 ± .015			
4	1/4 – 28 Thread Ø.250 Diameter ± .000-.002			
5	1/4 – 28 Thread .41 ± .015 Length			
6	Neck Ø.201 Diameter ± .005			
7	Neck .090 Wide ± .005			
8	Shoulder Ø.370 Diameter ± .005			
9	Shoulder .06 Width ± .015			
10	Taper 2.12 Length ± .015			
11	Taper Ø.31 Diameter ± .015			
12	Taper Ø.57 ± .015			
13	Front Groove in Knurls .125 Wide ± .005			
14	Front Groove in Knurls .312 Back ± .005			
15	Front Groove in Knurls Ø.575 Diameter ± .005			
16	Radius .28 ± .015			
17	Radius Diameter ± .010 .28			
18	**{Head}** 1.44 length ± .015			
19	Ø.50 Diameter ± .015			
20	Ø.250 Diameter ± .005			
21	.19 Radius ± .015			

(Continued)

	Print Dimensions and Tolerances	Student's Inspection	Instructor's Inspection	Instructor Comments
22	.06 Radius ± .015			
23	.09 Radius ± .015			
24	.50 Length in Center ± .015			
25	Ø.375 Spot-face ± .005			
26	#3 Drill .425 Deep ± .005			

Project Grade: ______________________

Plumb Bob

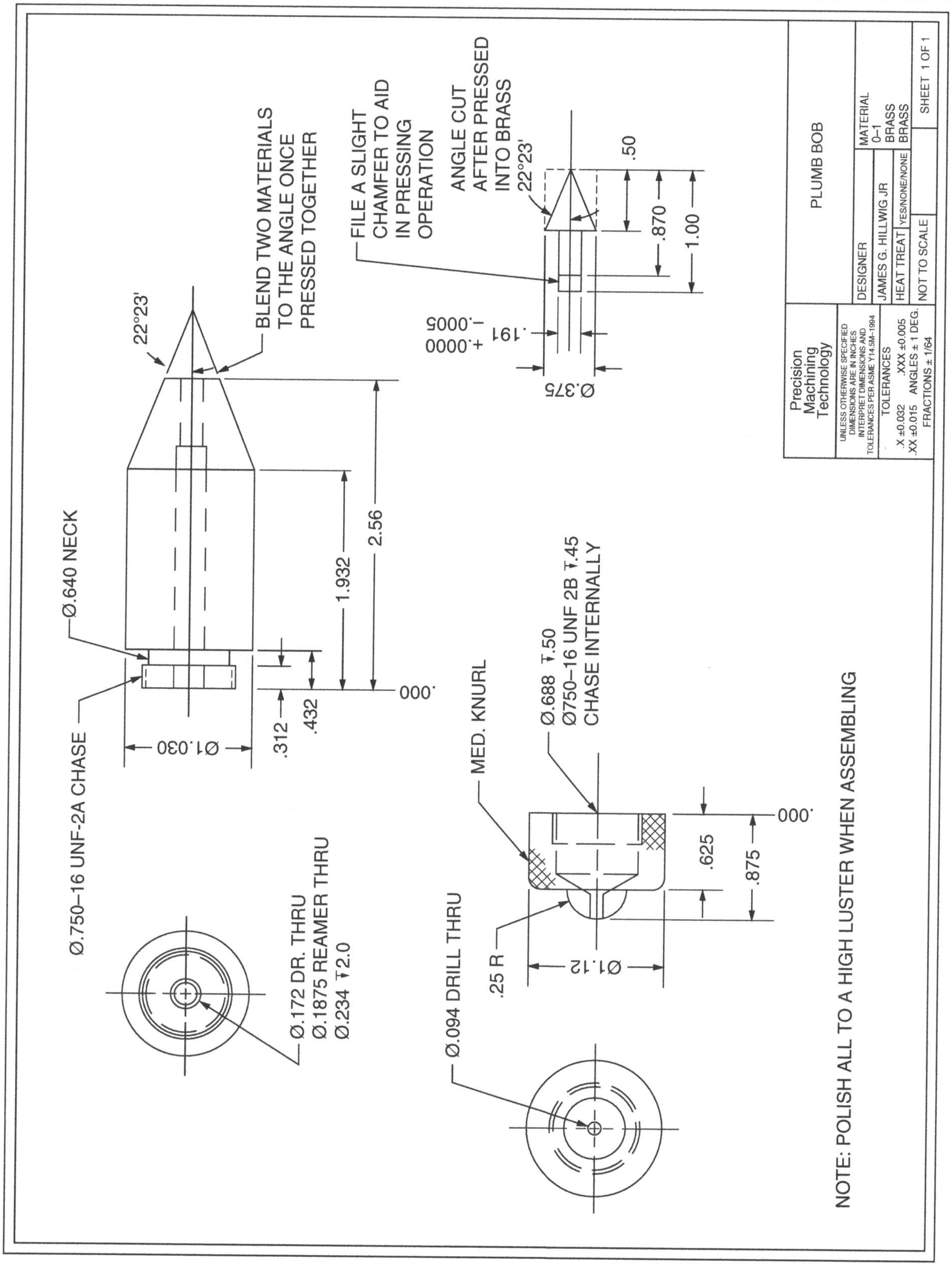
Ø.640 NECK
Ø.750–16 UNF-2A CHASE
Ø1.030
.312
.432
1.932
2.56
.000
22°23'
BLEND TWO MATERIALS TO THE ANGLE ONCE PRESSED TOGETHER
FILE A SLIGHT CHAMFER TO AID IN PRESSING OPERATION
ANGLE CUT AFTER PRESSED INTO BRASS
22°23'
.50
.870
1.00
.191 +.0000 −.0005
Ø.375
Ø.172 DR. THRU
Ø.1875 REAMER THRU
Ø.234 ↧2.0
MED. KNURL
Ø.688 ↧.50
Ø750–16 UNF 2B ↧.45
CHASE INTERNALLY
.625
.875
.000
.25 R
Ø1.12
Ø.094 DRILL THRU
NOTE: POLISH ALL TO A HIGH LUSTER WHEN ASSEMBLING
Precision Machining Technology
UNLESS OTHERWISE SPECIFIED DIMENSIONS ARE IN INCHES INTERPRET DIMENSIONS AND TOLERANCES PER ASME Y14.5M-1994
TOLERANCES
.X ±0.032 .XXX ±0.005
.XX ±0.015 ANGLES ± 1 DEG.
FRACTIONS ± 1/64
PLUMB BOB
DESIGNER
JAMES G. HILLWIG JR
MATERIAL
0-1
BRASS
BRASS
HEAT TREAT
YES/NONE/NONE
NOT TO SCALE
SHEET 1 OF 1

NIMS Skills Practiced

- Job Process Planning
- Turning Operations: Chucking

Equipment

Power saw, lathe with three-jaw and four-jaw chuck, bench grinder

Tools Needed

Marker
Various files
.625, 11/64, 3/32 drills
Various grits of sandpaper
Various types of turning, facing, grooving, forming, and threading tools
6″ steel rule
Center drill
.1875 reamer
Radius gages
File card
Knurling tool
Indicator w/magnetic base
Thread triangles

Material Needed

- 1-1/8″ diameter × 3-5/8″ long brass; 3/8″ × 1″ long drill rod (This needs to be heat treated 0-1; can be done with torches.)

Safety

1. Follow all shop safety rules given to you by your instructor and be sure to pass any safety tests given before using shop equipment.
2. Follow all safety rules for power equipment.
3. Be careful of sharp tools when using them.
4. Make sure the files have a handle on them.
5. Be sure all cutting tools are sharp and care is taken while handling them.
6. Be sure chuck is tightened and chuck wrench is never left in chuck.
7. Make sure stock is supported with center when necessary.

Order of Operations

BODY

1. Look over blueprint to determine tolerances, tools, and material needed.
2. Gather tools needed, figure out RPMs for required stock diameters and drills used, and write down their RPMs on sheet.
3. Measure and cut stock sizes needed.
4. Starting with the three-jaw chuck, take the 1-1/8″ brass, chuck with 2″ out of chuck, and face just enough off to clean up end and center drill. (Body)
5. Turn the 1.03 diameter back 1.875.
6. Turn the .750 diameter back .31.
7. Cut the .125 groove to .640 diameter .31 back.
8. Chase 3/4-16 thread (class 3A fit—check with triangles).

9. Turn end for end and hold in a four-jaw chuck (total indicator reading, or TIR, of .001 on the 1.032 diameter).
10. Center drill and knurl .875 from end on 1-1/8 diameter. (Cap)
11. Recheck TIR .001 and re-center drill if it was out.
12. Drill .562 deep from point of a 5/8 drill.
13. Now bore to .687 diameter and cut a .125 × .760 relief 1/2 back.
14. Chase a 3/4-16 internal thread.
15. Saw off cap .937 long.
16. Face part to 2.56 long. (Body)
17. Trig out angle and turn taper on the end; be careful not to go below .400 in diameter on small end.
18. Center drill and drill 11/64 through.
19. Ream .1875; there should be about a 1/32 chamfer left on hole from center drill.
20. Turn end for end in four-jaw chuck, TIR .001 on 1.03 diameter.
21. Screw on cap, face cap to clean up, center drill, and drill 3/32 through.
22. Turn .500 diameter back to allow .625 of knurls to remain.
23. Face .500 diameter to .250 long and form the .250 radius.

POINT

1. Use 3/8"-diameter tool steel.
2. Face and turn end for end and face to length.
3. Turn .191 diameter × 1/2 long.
4. File .125 long lead taper and then press into brass.
5. Hold in four-jaw chuck (TIR .001 on 1.03 diameter) and turn taper to point; this should include the brass by blending the angle seamlessly together.
6. Polish the tool steel to remove any machining marks.
7. Heat-treat tip.
8. Polish all over.

Student Name: ____________________ Date Submitted: __________

Class: __________ Total Hours on Job: __________

	Print Dimensions and Tolerances	Student's Inspection	Instructor's Inspection	Instructor Comments
1	Is the part free of burrs and does it meet finish requirements?			
2	Is the part properly identified?			
3	**(Part 1)** 2.56 ± .015			
4	Length .312 ± .005			
5	Length .432 ± .005			
6	Length 1.932 ± .005			
7	Diameter Ø 1.030 ± .005			
8	Diameter 3/4 – 16 Thread Ø .750 ± .000 − .005			
9	Neck .12 Length ± .015			
10	Neck Diameter Ø.640 ± .005			
11	Through Hole Ø.172 ± .005			
12	**{Part 2}** Length .875 ± .005			
13	Length .625 ± .005			
14	Depth .50 ± .015			
15	Diameter Ø 1.12 ± .015			
16	Radius .25 R ± .015			
17	Reamed Hole Ø.1875 ± .005			
18	Total Length 3.935 ± .015			
19	Are the Knurls to Points?			
20	Do the Angles Blend Between Tip and Body?			

Project Grade: __________

Threading Project

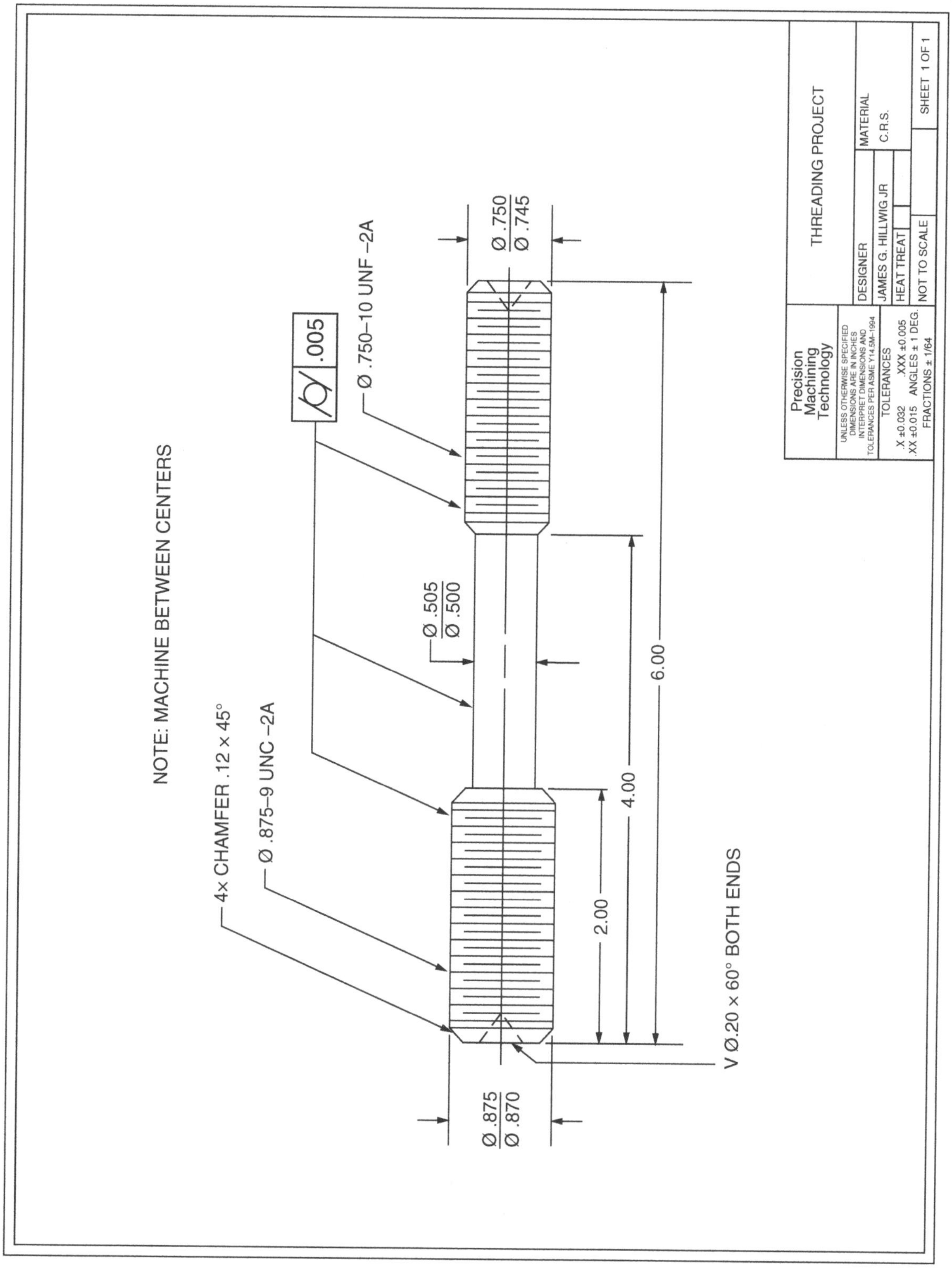

NIMS Skills Practiced

- Job Process Planning
- Turning Operations: Chucking

Equipment

Power saw, lathe, bench grinder

Tools Needed

6″ steel rule
Center drill
Various types of turning, facing, and threading tools
File card
Thread pitch gage
Layout dye
Various files
Go/no-go gages
Center gage

Material Needed

- Mild steel 1″ × 6-1/8″

Safety

1. Follow all shop safety rules given to you by your instructor and be sure to pass any safety tests given before using shop equipment.
2. Follow all safety rules for power equipment.
3. Be careful of sharp tools when using them.
4. Make sure the files have a handle on them.
5. Be sure all cutting tools are sharp and care is taken while handling them.
6. Be sure chuck is tightened and chuck wrench is never left in chuck.
7. Make sure stock is supported with center when necessary.

Order of Operations

1. Look over blueprint to determine tolerances, tools, and material needed.
2. Gather tools needed, figure out RPMs for required stock diameters and drills used, and write down their RPMs on sheet.
3. Measure and cut stock sizes needed.
4. Place stock in chuck, face, and center drill one end.
5. Turn stock in chuck, face stock to length, and center drill.
6. Extend stock 4-1/2″ from chuck face and support center in tailstock.
7. Turn .875 diameter 4″ long from end.
8. Turn part in chuck and extend stock 4-1/2″ from chuck face and support center in tailstock.
9. Turn .750 diameter 4″ long from end.
10. Now mark the part 2-1/8″ from end in tailstock.
11. Begin cutting the .500 diameter in the center without going past the 4″ length you just turned and without removing the 2-1/8″ mark you just made.
12. Turn part in chuck and extend stock 4-1/2″ from chuck face and support center in tailstock.
13. Complete the .500 diameter 4″ from end supported in the tailstock.
14. Polish the .500 diameter.

15. With your instructor's help, cut the 45° chamfers on each of the threaded sections as shown.
16. Place a lathe center in the chuck and re-cut the 60° taper on it so it runs true with the machine.
17. Place a lathe dog on one end of your part and support the other with the tailstock.
18. Paint the end you are going to thread with layout dye.
19. Set up your threading tool and machine for the proper threads per inch and touch off the part to set zeros.
20. Taking a light cut, engage the machine to thread; this should leave a light impression on the part you can check with the thread pitch gages.
21. If your thread setup is correct, finish your threads.
22. Check them along the way with thread micrometers or thread triangles.
23. Once to depth, check with go/no-go gage.
24. If gage does not fit, follow instructor's directions to pick up thread and correct the size.
25. Now turn part end for end, placing lathe dog on end you just threaded—place a penny or something over your threads where the screw touches so you do not damage the threads you just made, and place the other end in the tailstock.
26. Repeat steps 18 through 24.
27. Inspect part and turn in.

Student Name: ____________________ Date Submitted: ____________

Class: ____________ Total Hours on Job: ____________

	Print Dimensions and Tolerances	Student's Inspection	Instructor's Inspection	Instructor Comments
1	Is the part free of burrs and does it meet finish requirements?			
2	Is the part properly identified?			
3	Ø .875 $\pm$ .000 $-$ .005			
4	9 Pitch (Threads per Inch)			
5	Ø.505 Ø.500			
6	Ø.750 $\pm$ .000 $-$ .005			
7	6.00 Length $\pm$ .015			
8	10 Pitch (Threads per Inch)			
9	**(7/8 Length)** 2.00 $\pm$ .015			
10	**(.50 Length)** 2.00 $\pm$ .015			
11	**(3/4 Length)** 2.00 $\pm$ .015			
12	.12 $\times$ 45° Chamfer Ends and Middle			
13				
14				
15				

Project Grade: ____________

Machining Level 1 Preparation Projects—Turning Between Centers

Jack Screws

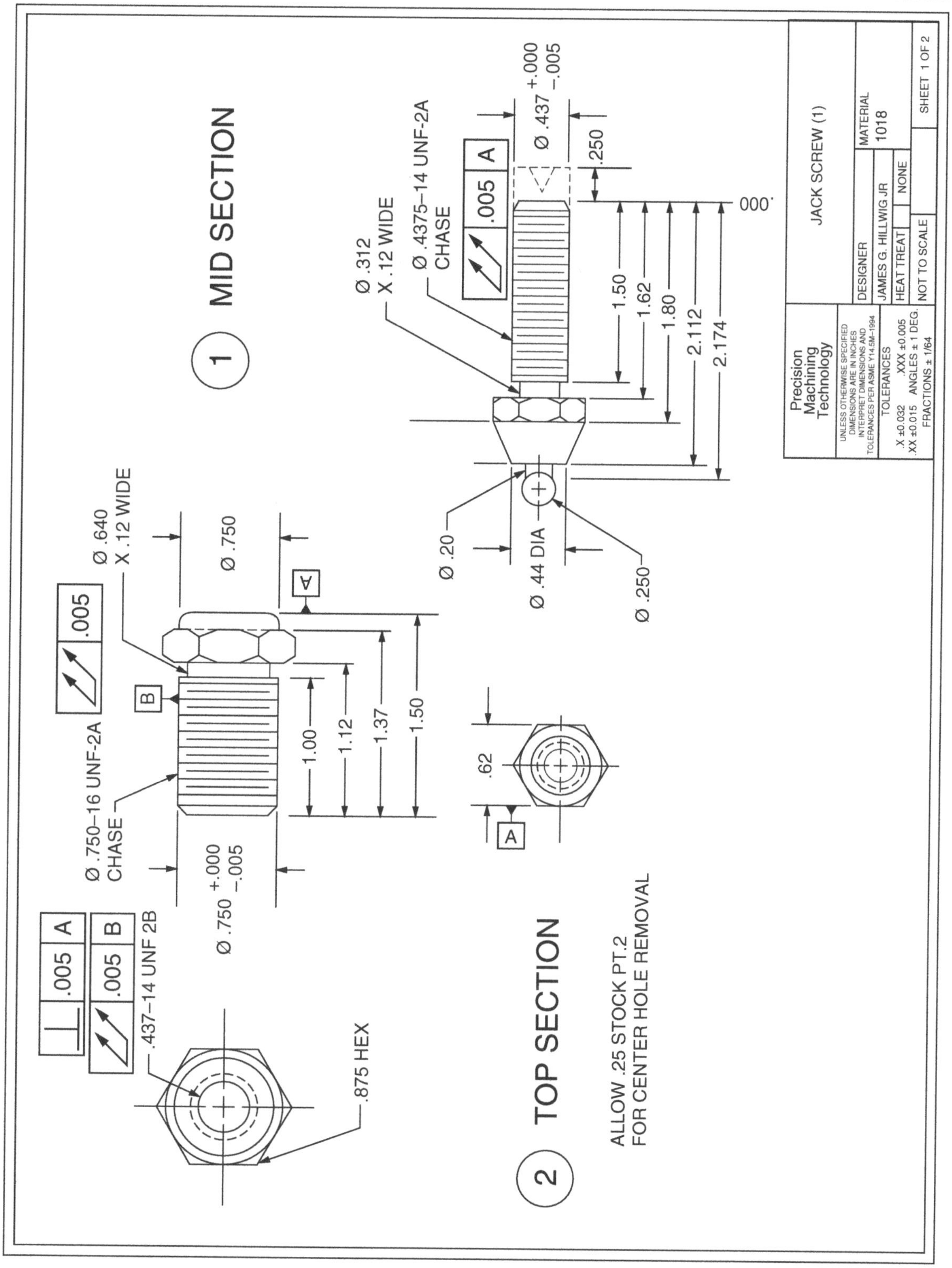

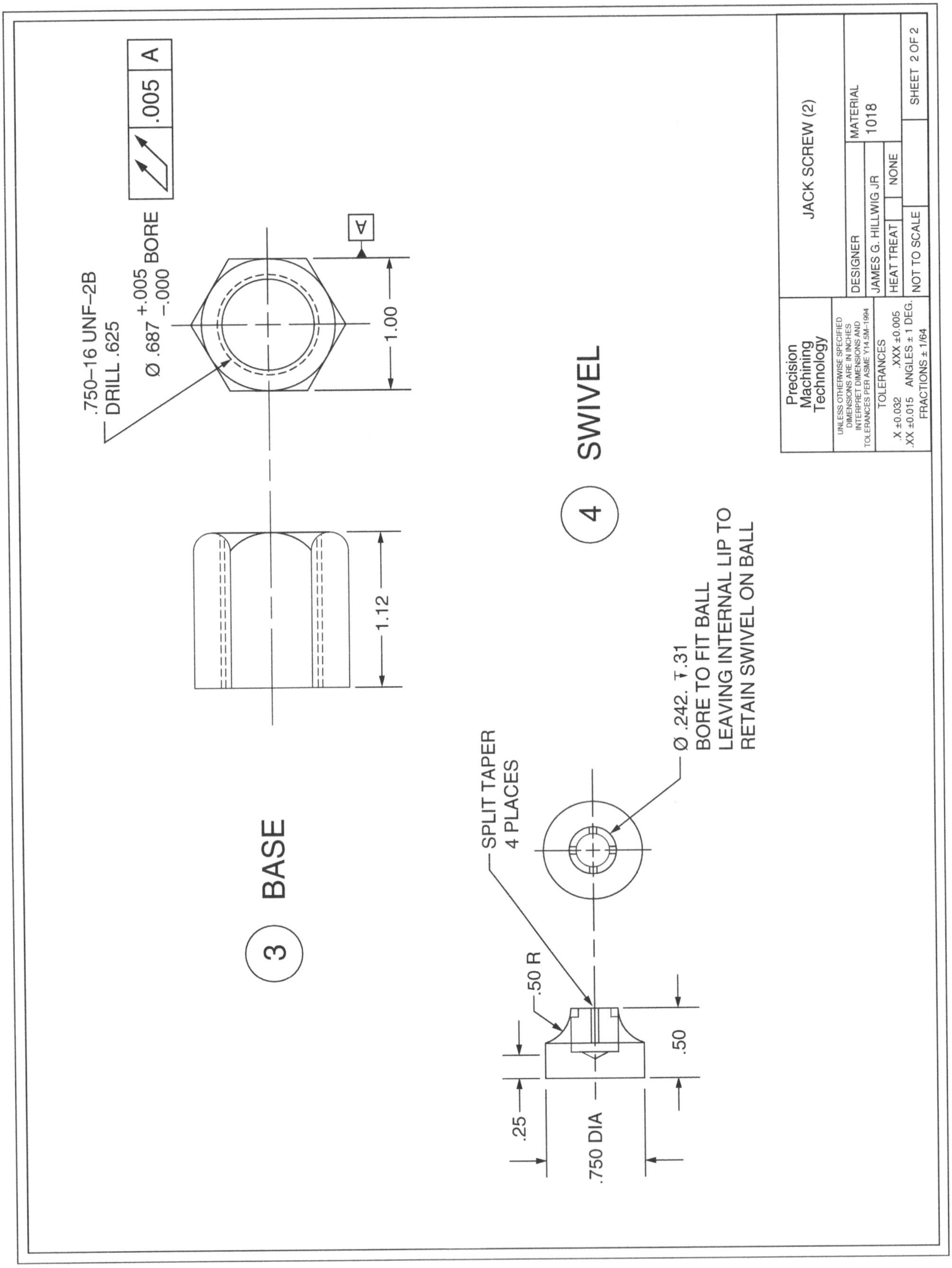
3 BASE
.50 R
SPLIT TAPER 4 PLACES
.25
.50
.750 DIA
Ø .242. ⊽ .31
BORE TO FIT BALL
LEAVING INTERNAL LIP TO
RETAIN SWIVEL ON BALL
4 SWIVEL
.750–16 UNF–2B
DRILL .625
Ø .687 +.005 -.000 BORE
.005 A
A
1.00
1.12
Precision Machining Technology
UNLESS OTHERWISE SPECIFIED DIMENSIONS ARE IN INCHES INTERPRET DIMENSIONS AND TOLERANCES PER ASME Y14.5M-1994
TOLERANCES
.X ±0.032 .XXX ±0.005
.XX ±0.015 ANGLES ± 1 DEG.
FRACTIONS ± 1/64
JACK SCREW (2)
DESIGNER
JAMES G. HILLWIG JR
MATERIAL
1018
HEAT TREAT
NONE
NOT TO SCALE
SHEET 2 OF 2

NIMS Skills Practiced

- Job Process Planning
- Turning Operations: Chucking
- Turning Operations: Between Centers

Equipment

Power saw, lathe, bench vise, bench grinder

Tools Needed

Marker
Files
Radius gages
Hacksaw
Various types of turning, facing, grooving, forming, boring, and threading tools
6″ steel rule
Center drill
Thread triangles
Lathe center
Large angle plate
Half nuts
File card
Various drills
Various taps
Lathe dog
Shim stock
Compound clamps

Material Needed

- HEX stock (CRS):
  - 1 @ 5/8″ × 5-3/8″ long (Part #2 Top Section)
  - 1 @ 7/8″ × 3-3/8″ long (Part #1 Mid Section)
  - 2 @ 1″ × 1-1/8″ long (Part #3 Base)
- ROUND Stock (CRS):
  - 1 @ 3/4″ × 2-1/4″ long (Part #4 Swivel)

Safety

1. Follow all shop safety rules given to you by your instructor and be sure to pass any safety tests given before using shop equipment.
2. Follow all safety rules for power equipment.
3. Be careful of sharp tools when using them.
4. Make sure the files have a handle on them.
5. Be sure all cutting tools are sharp and care is taken while handling them.
6. Be sure chuck is tightened and chuck wrench is never left in chuck.
7. Make sure stock is supported with center when necessary.

Order of Operations

YOU ARE MAKING TWO SETS OF JACK SCREWS

1. Look over blueprint to determine tolerances, tools, and material needed.
2. Gather tools needed, figure out RPMs for required stock diameters and drills used, and write down their RPMs on sheet.
3. Measure and cut stock sizes needed.
4. Parts #1 and #2 are made by double ending.
5. Do similar operations on parts #1 and #2 together.

6. Face and center drill ends of both pieces.
7. Part #1: between centers turn .750 diameter × 1.12 long on each end.
8. Part #2: between centers turn .437 diameter × 1.75 long on each end.
9. Part #1: hold in chuck, bring live center into centered-drilled hole, and cut .640 diameter × .12 wide neck 1.12 back from end. Flip end for end and repeat.
10. Part #2: hold in chuck, bring live center into centered-drilled hole, and cut .312 diameter × .12 wide neck 1.75 back from end. Flip end for end and repeat.
11. Part #1: between centers, chase 3/4-16 thread, class 3A fit, on each end. (Use thread triangles to check size; use half nuts to protect threads where the lathe dog touches.)
12. Part #2: chase 7/16-14 thread on both ends same as part #1.
13. Saw both parts #1 and #2 in half.
14. Part #3: mark one flat on each part with dye or a marker and always keep that flat on jaw #1. It is important to do this and use the same machine or the part will not run true enough.
15. Face both ends of both pieces and take to size 1.12 long.
16. Center drill, pilot drill, and then drill 5/8" hole through. Do both pieces.
17. Bore both pieces to .687.
18. Chase 3/4-16 internal threads to fit part #1 in both pieces.
19. Crown one end of each #3 part as shown on print.
20. Screw part #1 into part #3; make sure marked flat is lined up with jaw #1.
21. Face part #1 to length (.37" past the crowned end of part #3 or unscrew and check that it is 1.50 long).
22. Repeat on the other parts #1 and #3.
23. Turn .750 diameter by .12 long and file the 1/16" radius on both #1 pieces.
24. With marked flat on jaw #1, center drill, drill, and tap both #1 pieces.
25. Screw part #2 into part #1, which is screwed into part #3, keeping the marked flat on jaw #1.
26. Face the 5/8 hex to .812 long from part #1 and turn 1/4 diameter to .260 diameter by .312 long. Talk to your instructor about power tapping on the lathe.
27. Form ball and neck, then polish to .250 diameter.
28. Trig out angle to find the taper—same formula as used for Reversible Pocket Scribe and Plumb Bob.
29. Machine angle to 5/16 long on both pieces. Do NOT cut off corners of the hex.
30. Remove from chuck; make sure the pieces are screwed together tight.
31. Turn end for end and return to chuck with the marked flat on jaw #1.
32. Now face the bottom to a uniform appearance and the proper length; the center-drilled hole should be gone in part #2.
33. Chuck 3/4 stock in lathe; face and center drill (swivel).
34. Drill approximately .240 diameter 5/16 deep, then bore to .255, leaving about a .03 to .05 lip at the opening of the hole; this will help keep the swivel on the ball.
35. Mark 3/8 diameter on face and mark back 1/4.
36. Form the radius with a radius tool, file, polish, and check with a radius gage.
37. Turn end for end and repeat process.
38. Cut to length using a cutoff tool and deburr.
39. Hacksaw splits while holding in vise.
40. Press swivel on ball.

Student Name: ______________________ Date Submitted: ____________

Class: ____________ Total Hours on Job: ____________

	Print Dimensions and Tolerances	Student's Inspection	Instructor's Inspection	Instructor Comments
1	Is the part free of burrs and does it meet finish requirements?			
2	Is the part properly identified?			
3	**BASE Part #3** 1.12 ± .015			
4	1.00 ± .015			
5	.687 ± .005/−.000			
6	.750-16 THD Drill .625			
7	**SWIVEL** Ø.750 ± .005			
8	.50R ± .015			
9	.25 ± .015			
10	Ø.25 Drill .31 DP			
11	**MID SECTION Part #1** .62 ± .015			
12	Ø.44 ± .015			
13	Ø.250 ± .005			
14	.062 ± .005			
15	.18 ± .015			
16	.312 ± .005			
17	Ø.312 × .12			
18	1.50 ± .015			
19	Ø.437 ± .000 − .005			
20	Ø.4375 − 14			

(Continued)

	Print Dimensions and Tolerances	Student's Inspection	Instructor's Inspection	Instructor Comments
21	**TOP SECTION Part #2** .875 HEX ± .005			
22	Ø.437 – 14 TAP			
23	Ø.750-16 THD. CHASE			
24	Ø.640 ± .005			
25	Ø.750 ± 000/–.005			
26	1.12 ± .015			
27	1.50 ± .015			
28	.12 NECK ± .015			
29	.25 ± .015			
30	.12 ± .015			
31	Ø.750 ± .015			
32	.37 ± .015			

Project Grade: ____________________

Flat Scribe

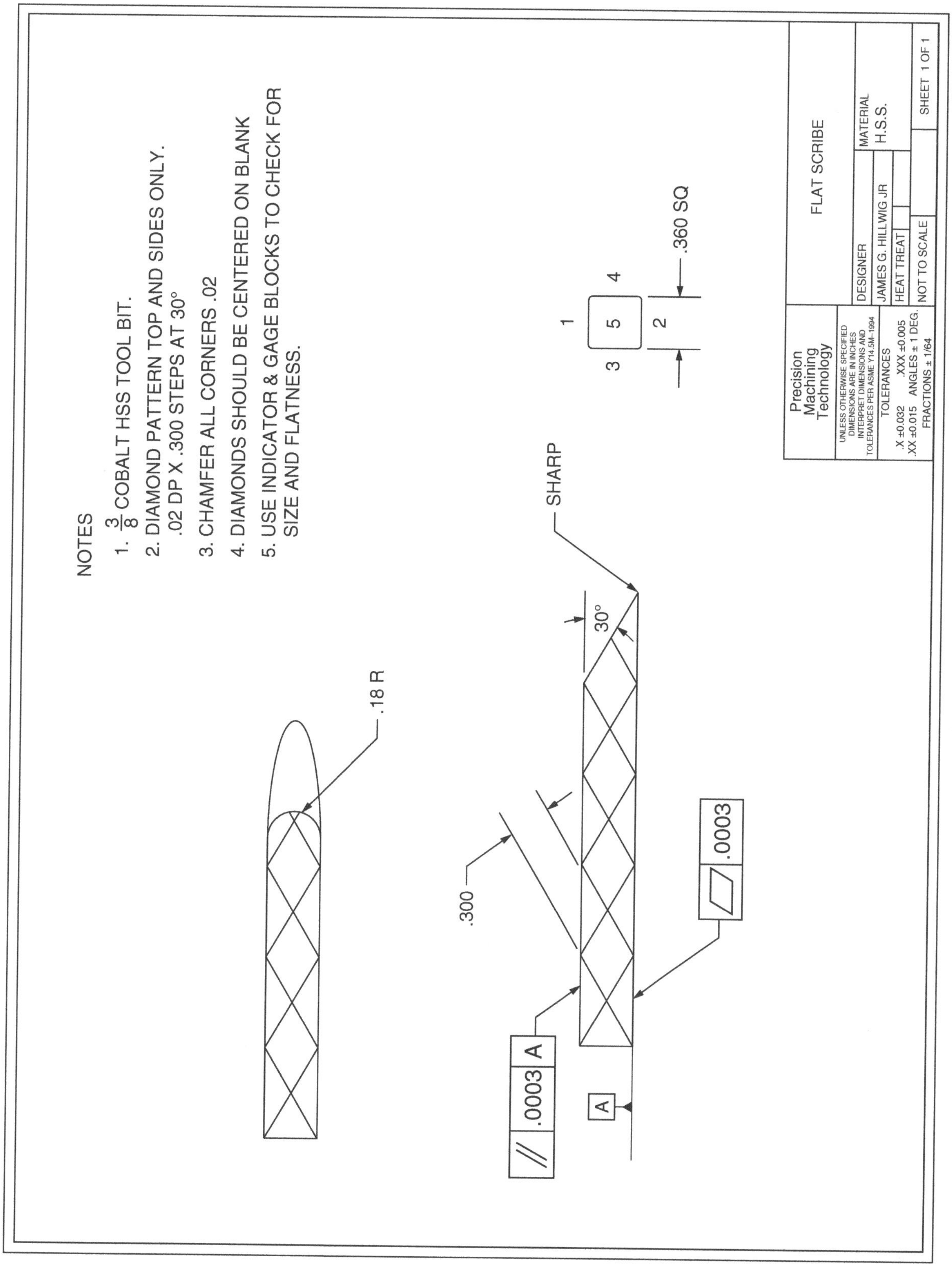
NOTES
1. 3/8 COBALT HSS TOOL BIT.
2. DIAMOND PATTERN TOP AND SIDES ONLY. .02 DP X .300 STEPS AT 30°
3. CHAMFER ALL CORNERS .02
4. DIAMONDS SHOULD BE CENTERED ON BLANK
5. USE INDICATOR & GAGE BLOCKS TO CHECK FOR SIZE AND FLATNESS.
.18 R
SHARP
30°
.300
.0003
.0003 A
A
.360 SQ
1
2
3
4
5
FLAT SCRIBE
Precision Machining Technology
UNLESS OTHERWISE SPECIFIED DIMENSIONS ARE IN INCHES INTERPRET DIMENSIONS AND TOLERANCES PER ASME Y14.5M-1994
TOLERANCES
.X ±0.032 .XXX ±0.005
.XX ±0.015 ANGLES ± 1 DEG.
FRACTIONS ± 1/64
DESIGNER
JAMES G. HILLWIG JR
MATERIAL
H.S.S.
HEAT TREAT
NOT TO SCALE
SHEET 1 OF 1

NIMS Skills Practiced

- Job Process Planning
- Surface Grinding and Safety
- Squaring Up a Part on Surface Grinder

Equipment

Surface grinder

Tools Needed

Diamond dresser
Grinder vise
Demagnetizer
Surface plate
Grinder vise parallels
Cutoff wheel
Indicator and base
Comparator
Grinding wheel
Radius dresser
Gage blocks
Chuck stone

Material Needed

- 3/8″ × 3/8″ HSS square tool bit

Safety

1. Follow all shop safety rules given to you by your instructor and be sure to pass any safety tests given before using shop equipment.
2. Follow all safety rules for power equipment.
3. Be careful of sharp tools when using them.
4. Ring grinding wheels before using them.
5. Make sure magnetic chuck is on when surface grinding.

Order of Operations

1. Look over blueprint to determine tolerances, tools, and material needed.
2. Measure the tool bit blank to be sure it is 3/8″ × 3/8″.
3. Deburr all over.

MAKE SURE MAGNETIC CHUCK IS ON WHEN SURFACE GRINDING AND FOLLOW ALL SAFETY RULES!

4. Check the magnetic chuck for raised burrs, and stone if necessary.
5. Mount and dress a 46-H grinding wheel.
6. Place the tool bit on the magnetic chuck at about a 40° angle to the back rail (to cover more magnetic strips), and place square blocking on both sides of the tool bit to keep it in place.
7. Be cautious of the tool bit lifting from chuck while grinding. (It will make a shearing noise if it begins to lift.)
8. Turn the magnetic chuck on.
9. Grind surface #1 until it cleans up completely.
10. Deburr the edges of side #1. BE CAREFUL: the burrs are very sharp.
11. Place side #1 on the chuck so you can grind side #2. Follow the same setup as for side #1. You can make it to size.
12. Deburr and check for flatness and size.

13. Place tool bit in grinder vise on parallels with ground sides against jaws. Leave one edge sticking out of side of vise so that you can turn vise on its side.
14. Grind top of part (side #3) to clean up and turn vise on side (side #5) without removing or disturbing part while in vise just to clean it up.
15. Remove from vise, deburr all edges, and place side #3 on magnetic chuck with square blocking on each side and grind side to size.
16. Deburr and check for flatness and size.
17. Place tool bit in grinder vise on a 30° flat angle and grind the last edge to a point.
18. Be careful not to create excessive heat while grinding angle; do not let the part change colors on you.
19. Remove the vise from the chuck. Using the radius dresser, with your instructor's help, dress a .18 radius on the wheel.
20. Place grinder vise back on chuck against the back rail and locate the center of the tool bit with the center of the radius.
21. Grind tool bit until you get a complete radius on the 30° angle.
22. Be careful not to create excessive heat while grinding radius; do not let the part change colors on you.
23. Remove vise from chuck, remove tool bit from vise, and deburr.
24. Using the same 30° angle, place the angle along the back rail and the tool bit against the angle, top up. Be sure to block the other side of the tool bit.
25. Place a cutoff wheel on surface grinder.
26. Find the corner of the tool bit and move over the part .300 plus 1/2 the thickness of the cutoff wheel. Set zero.
27. Touch-off the top of the part and set your zero. Slowly feed down .020 deep while moving table right to left.
28. Raise the wheel past zero, move table in .300, and slowly feed down .020 deep while moving the table right to left.
29. Repeat this process until you reach the other end.
30. Repeat this process on the two sides, and then turn the 30° angle to make the "X" complete. You should still be able to return to the same original zero for the first line.
31. As the wheel wears down, you may need to reset your downfeed to zero.
32. DO NOT grind "X" on bottom of part.
33. Grind a .020-by-45° chamfer on all edges.
34. Inspect the part to print dimensions; record on inspection sheet and turn in.

Student Name: ______________________ Date Submitted: ____________

Class: ______________ Total Hours on Job: ____________

	Print Dimensions and Tolerances	Student's Inspection	Instructor's Inspection	Instructor Comments
1	Is the part free of burrs and does it meet finish requirements?			
2	Is the part properly identified?			
3	Are sides .360 ± .005?			
4	Is the part flat within ± .0003?			
5	Is the angle 30° ± 1°?			
6	Is the radius .18 ± .015?			
7	Are the "X's" .300 ± .005 apart?			
8	Are the "X's" to .02 ± .015?			
9	Are the "X's" centered in the part?			
10	Are the chamfers on all corners .02 deep ± .015?			
11				
12				
13				
14				
15				

Project Grade: ____________

Compound Clamp

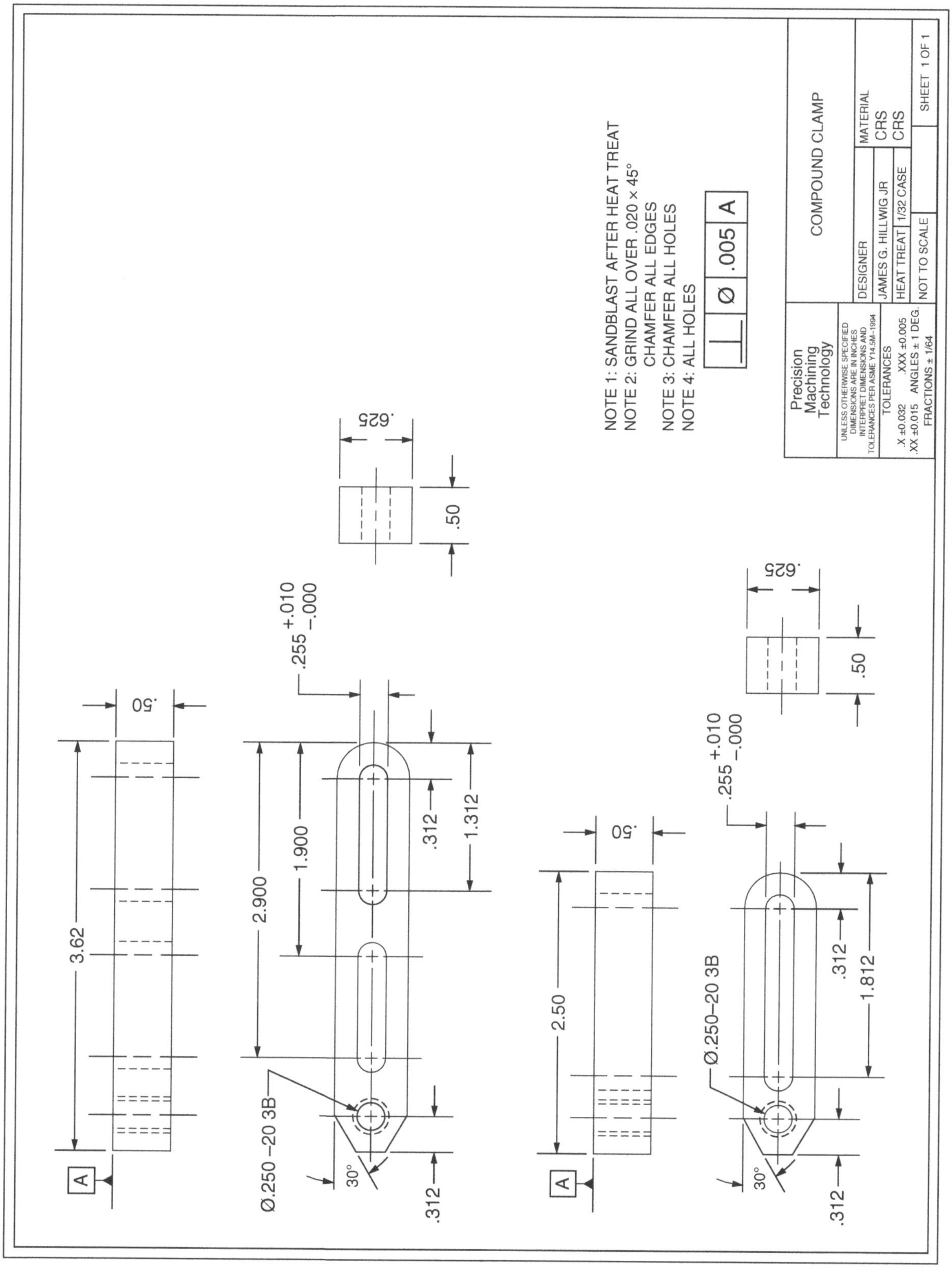

NIMS Skills Practiced

- Job Process Planning
- Layout
- Drilling and Tapping
- Surface Grinding and Safety
- CNC Milling and Safety
- CNC Setup and Operation
- CNC Programming

Equipment

Power saw, CNC milling machine, milling machine, surface grinder

Tools Needed

Scribe
Dead blow hammer
Various drill bits
90° countersink
Edge finder
Machinist square
Magnetic sine plate
6″ steel rule
File
1/4-20 tap
Various sizes of endmills
Center drill
Grinding wheel
Layout dye
File card
Stamps
Parallels
Vise stop
Grinder vise

Material Needed

- 4 @ 3/4″ square × 2-5/8″ long CRS (Part #2)
- 2 @ 3/4″ square × 3-3/4″ long CRS (Part #1)

Safety

1. Follow all shop safety rules given to you by your instructor and be sure to pass any safety tests given before using shop equipment.
2. Follow all safety rules for power equipment.
3. Be careful of sharp tools when using them.
4. Make sure the file has a handle on it.
5. Be sure all cutting tools are sharp and care is taken while handling them.
6. Follow all CNC rules and safety guidelines for operation and programming.
7. Ring grinding wheels before using them.
8. Make sure magnetic chuck is on when surface grinding.

Order of Operations

MILL

1. Look over blueprint to determine tolerances, tools, and materials needed.
2. Gather tools needed, figure out RPMs for required tools, and write down their RPMs on sheet.
3. Measure, mark, and cut stock.

4. Deburr all over.
5. Write program for cutting out parts. Use the top view for the program.
6. Load tools in correct pockets on CNC mill.
7. Place part in mill vise on parallels, leaving .650 sticking out of the top of the vise (or whatever your instructor requires), and tap down with dead blow hammer.
8. Place vise stop against edge of part, being sure it is out of the tool path.
9. Load program.
10. Edge find part and set work offsets.
11. Set tool length offsets.
12. Dry run program or single block (follow instructor's requirements).
13. Run program to cut part out.
    a. Contour outside of part.
    b. Center drill holes, tapped hole, and ends of slots.
    c. Drill tapped holes through with #7 drill.
    d. Mill slots.
    e. Repeat with remaining parts of same size.
    f. Parts of different length need a new program written.
14. The remaining material left in the vise is called the hat; you will mill that off on the manual mill and then cut your name pocket in the side, right behind the angle.
15. Set part in mill and mill to thickness.
16. Mill a name pocket centered on the 1/2″ dimension just behind the angle ways .030 deep, 1/4″ wide, and 3/8″ long.
17. Deburr, place on anvil, and, using steel stamps, stamp your name in pocket.

This part can be heat-treated if your budget permits, but it is not necessary.

SURFACE GRINDER

MAKE SURE MAGNETIC CHUCK IS ON WHEN SURFACE GRINDING!

18. Surface grind all sides square and parallel. Deburr after each side is ground.
19. Now check sizes, hole location, and slot; grind them into location and part to size. Be sure to deburr and block part when needed.
20. Using vise and magnetic sine bar, calculate gage block height, set magnetic sine bar to height, place part in grinder vise, place on magnetic sine bar, TURN ON MAGNETS, and grind angle.
21. Using a sine bar and roll dimensions, check the angle. Have your instructor help you.
22. Lay flat on grinder chuck and grind a .030-by-45° chamfer on all edges.
23. Evaluate and turn in.

Student Name: ______________________ Date Submitted: ____________

Class: ______________ Total Hours on Job: ____________

	Print Dimensions and Tolerances	Student's Inspection	Instructor's Inspection	Instructor Comments
	Part #1—Clamp			
1	Is the part free of burrs and does it meet grind finish requirements?			
2	Is the part properly identified?			
3	**{Length}** 2.900 ± .005			
4	**{Length}** 1.900 ± .005			
5	**{Length}** 1.312 ± .005			
6	**{Radius}** .312 ± .005			
7	**{Length}** 3.62 ± .015			
8	**{Height}** .625 ± .005			
9	**{Width}** .50 ± .015			
10	**{Width}** .255 +.010/−.000			
11	30° Angle ± 1°			
12	**{¼-20 3A Thread}** Good Form?			
	Part #2—Clamp			
13	Is the part free of burrs and does it meet finish requirements?			
14	Is the part properly identified?			
15	**{Length}** 2.50 ± .015			
16	**{Height}** .625 ± .005			
17	**{Width}** .50 ± .015			
18	30° Angle ± 1°			
19	**{¼-20 3A Thread}** Good Form?			

(Continued)

	Print Dimensions and Tolerances	Student's Inspection	Instructor's Inspection	Instructor Comments
20	**{THD. Position ± .015}** Ø.250 – 20			
21	**{Width}** .255 +.010/−.000			
22	**{Length}** 1.812 ± .005			
23	**{Radius}** .3125 ± .005			

Project Grade: ____________________

Key Chain

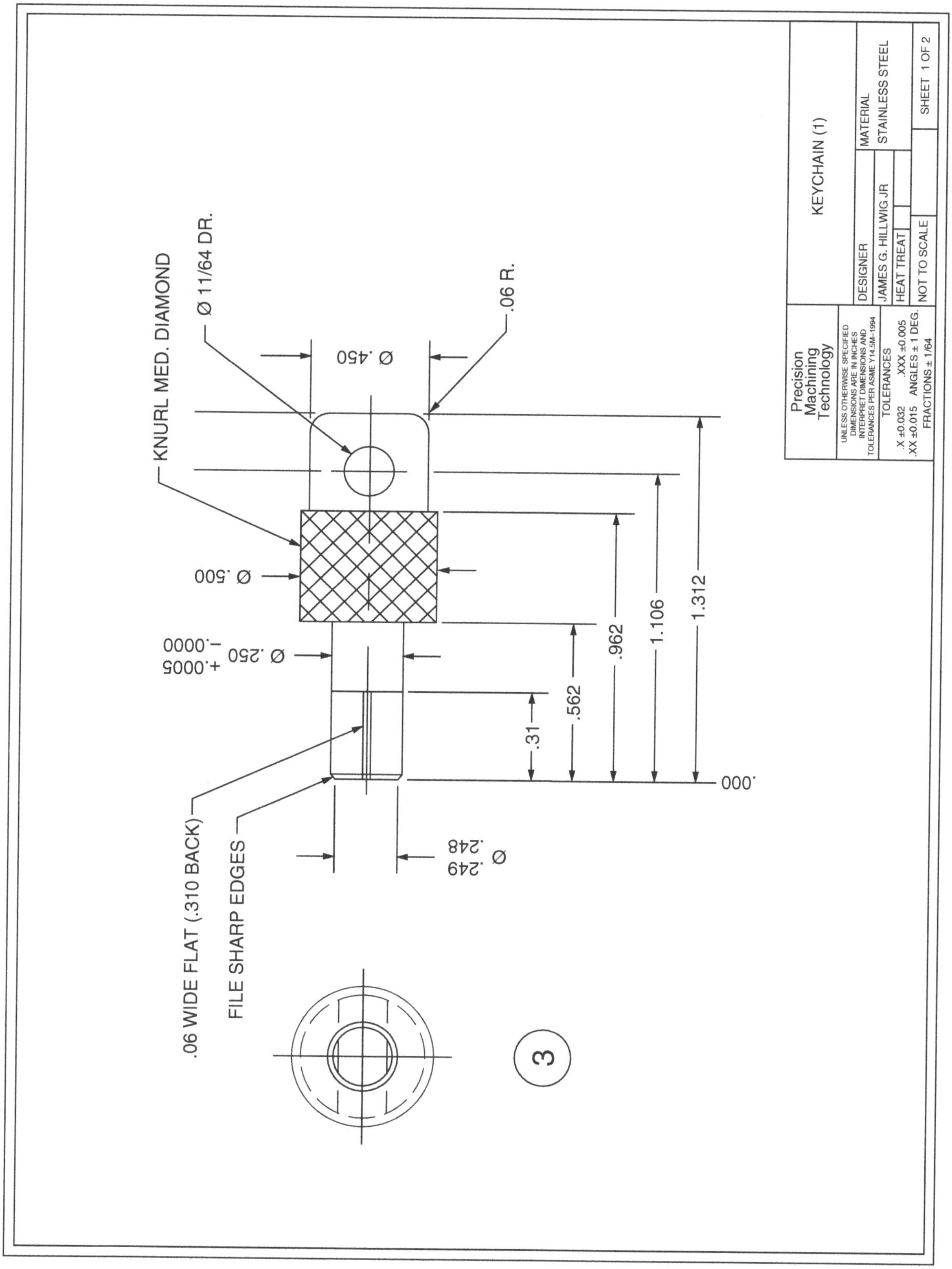
KNURL MED. DIAMOND
Ø 11/64 DR.
.06 R.
Ø .450
Ø .500
Ø .250 +.0005 −.0000
.06 WIDE FLAT (.310 BACK)
FILE SHARP EDGES
Ø .249 .248
.31
.562
.962
1.106
1.312
.000
3
KEYCHAIN (1)
Precision Machining Technology
UNLESS OTHERWISE SPECIFIED DIMENSIONS ARE IN INCHES INTERPRET DIMENSIONS AND TOLERANCES PER ASME Y14.5M-1994
TOLERANCES
.X ±0.032 .XXX ±0.005
.XX ±0.015 ANGLES ± 1 DEG.
FRACTIONS ± 1/64
DESIGNER
JAMES G. HILLWIG JR
MATERIAL
STAINLESS STEEL
HEAT TREAT
NOT TO SCALE
SHEET 1 OF 2

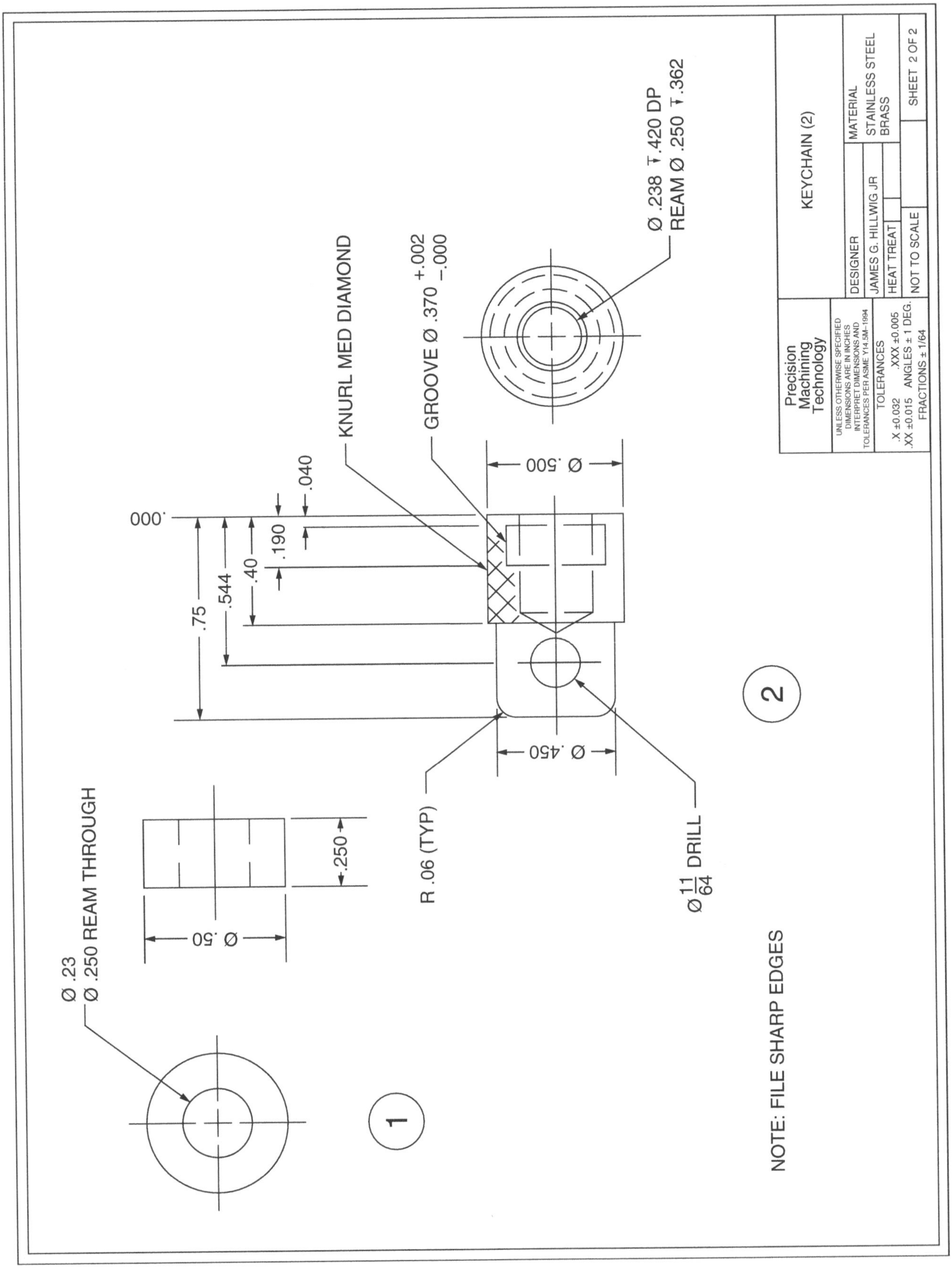
Ø .23
Ø .250 REAM THROUGH
Ø .50
.250
1
NOTE: FILE SHARP EDGES
.000
.75
.544
.40
.190
.040
KNURL MED DIAMOND
GROOVE Ø .370 +.002 −.000
Ø .500
Ø .450
R .06 (TYP)
Ø 11/64 DRILL
2
Ø .238 ⊤.420 DP
REAM Ø .250 ⊤.362
Precision Machining Technology
UNLESS OTHERWISE SPECIFIED DIMENSIONS ARE IN INCHES INTERPRET DIMENSIONS AND TOLERANCES PER ASME Y14.5M–1994
TOLERANCES
.X ±0.032 .XXX ±0.005
.XX ±0.015 ANGLES ± 1 DEG.
FRACTIONS ± 1/64
KEYCHAIN (2)
DESIGNER
JAMES G. HILLWIG JR
HEAT TREAT
NOT TO SCALE
MATERIAL
STAINLESS STEEL
BRASS
SHEET 2 OF 2

NIMS Skills Practiced

- Job Process Planning
- Turning Operations: CNC Chucking

Equipment

Power saw, CNC lathe, mill, bench grinder

Tools Needed

Marker
Various files
Various drills
Various types of turning, facing, boring, and forming tools
6″ steel rule
Center drill
.260 reamer
File card
Knurling tool
Radius gages

Material Needed

- 1/2″ diameter × .250″ long brass (ask your instructor if you should use a longer piece to hold safely in chuck, and then cut it off)
- 1 @ 1/2″ diameter × 1.″ long stainless steel
- 1 @ 1/2″ diameter × 1.5.″ long stainless steel
- 2 @ 1/4″ O-rings (insert them in the bored groove; this will hold both pieces together)

Safety

1. Follow all shop safety rules given to you by your instructor and be sure to pass any safety tests given before using shop equipment.
2. Follow all safety rules for power equipment.
3. Be careful of sharp tools when using them.
4. Make sure the files have a handle on them.
5. Be sure all cutting tools are sharp and care is taken while handling them.
6. Follow all CNC rules and safety guidelines for operation and programming.
7. Be sure chuck is tightened and chuck wrench is never left in chuck.
8. Make sure stock is supported with center when necessary.

Order of Operations

PART #1

1. Look over blueprint to determine tolerances, tools, and materials needed.
2. Gather tools needed, figure out RPMs for required stock diameters and drills used, and write down their RPMs on sheet.
3. Measure and cut stock sizes needed.
4. Write programs for cutting out parts.
5. Load tools in turret on CNC lathe.
6. Place part #1 in chuck with about 1/2″ sticking out.

7. Load program.
8. Set tool length offsets.
9. Dry run program or single block (follow instructor's requirements).
10. Run program to cut part out.
    a. Face part.
    b. Center drill hole.
    c. Drill and ream.
    d. Cut off part.

PART #2

1. Place part #2 in chuck with about 1" sticking out.
2. Load program for part #2.
3. Set tool length offsets.
4. Dry run program or single block (follow instructor's requirements).
5. Run program to cut part out.
    a. Face part.
    b. Knurl part.
    c. Center drill part.
    d. Drill .238-diameter hole.
    e. Ream hole to .250 diameter.
    f. Bore internal groove.
    g. Cut off part.
6. Place part back in chuck, hole first.
7. Set tool length offsets.
8. Dry run program or single block (follow instructor's requirements).
9. Run program to cut .450 diameter by .350 long with .06 radius.

PART #3

1. Place part #3 in chuck with about 1.125" sticking out.
2. Load program for part #3.
3. Set tool length offsets.
4. Dry run program or single block (follow instructor's requirements).
5. Run program to cut part out.
    a. Face part.
    b. Knurl part.
    c. Turn .250 diameter by .562 long.
    d. Turn .248 to .249 by .31 long.
    e. Chamfer while cutting off part.
6. Place part back in chuck, pin side first.
7. Set tool length offsets.
8. Dry run program or single block (follow instructor's requirements).
9. Run program to cut .450 diameter by .350 long with .06 radius.
10. Set part #2 in mill, center drill, and drill 11/64-diameter hole.
11. Set part #3 in mill, center drill, and drill 11/64-diameter hole.
12. Mill or file .06 wide flat .310 as shown.

Student Name: ______________________ Date Submitted: ____________

Class: ____________ Total Hours on Job: ____________

	Print Dimensions and Tolerances	Student's Inspection	Instructor's Inspection	Instructor Comments
1	Is the part free of burrs and does it meet finish requirements?			
2	Is the part properly identified?			
3	**{Part #1 Brass}** .50 Diameter ± .015			
4	(Hole Diameter) .250 ± .005			
5	(Length) .250 ± .005			
6	**{Part #2 Stainless Steel}** .500 Diameter ± .005			
7	(Length) .75 ± .015			
8	(Diameter) .450 ± .005			
9	(Radius) .06 ± .015			
10	(Length) .350 ± .005			
11	(Internal Hole Diameter) .250 ± .005			
12	(Internal Hole Length) .420 ± .005			
13	(Internal Groove Diameter) .370 ± .002 -.000			
14	(Internal Groove Width) .150 ± .005			
15	(Internal Groove Lip) .040 ± .005			
16	(Drilled Hole Diameter) 11/64 ± 1/64			
17	(Drilled Hole Location) .206 ± .005			
18	**{Part #3 Stainless Steel}** .500 Diameter ± .005			
19	(Length) 1.312 ± .005			
20	(Diameter) .450 ± .005			
21	(Radius) .06 ± .015			

(Continued)

	Print Dimensions and Tolerances	Student's Inspection	Instructor's Inspection	Instructor Comments
22	(Length) .350 ± .005			
23	(Knurl Length) .40 ± .015			
24	(Internal Hole Length) .420 ± .005			
25	(Pin Length) .562 ± .005			
26	(Pin Diameter) .249 to .248			
27	(Pin Length with Flats) .31 ± .015			
28	(Pin Diameter) .250 ± .001 - .000			
29	(Pin Flats) .06 wide ± .015			
30	(Drilled Hole Diameter) 11/64 ± 1/64			
31	(Drilled Hole Location) .206 ± .005			